THE STORY OF
THE UNIVERSE
IN 100 STARS

The
STORY
of the
UNIVERSE
in
100 STARS

FLORIAN FREISTETTER

THE EXPERIMENT

NEW YORK

Originally published in Germany as *Eine Geschichte des Universums in 100 Sternen* by Carl Hanser Verlag, Munich, in 2019. First published in the UK as *A History of the Universe in 100 Stars* by Quercus Editions Ltd., a Hachette UK company, in 2021. First published in North America in revised form by The Experiment, LLC, in 2021.

The Experiment, LLC
220 East 23rd Street, Suite 600
New York, NY 10001-4658
theexperimentpublishing.com

THE EXPERIMENT and its colophon are registered trademarks of The Experiment, LLC. Many of the designations used by manufacturers and sellers to distinguish their products are claimed as trademarks. Where those designations appear in this book and The Experiment was aware of a trademark claim, the designations have been capitalized.

The Experiment's books are available at special discounts when purchased in bulk for premiums and sales promotions as well as for fundraising or educational use. For details, contact us at info@theexperimentpublishing.com.

Library of Congress Cataloging-in-Publication Data

Names: Freistetter, Florian, 1977- author. | Ipsen, Gesche, translator.
Title: The story of the universe in 100 stars / Florian Freistetter ; [translated by] Gesche Ipsen.
Other titles: Geschichte des Universums in 100 Sternen. English
Description: New York, NY : The Experiment, 2021. | "Originally published in Germany as Eine Geschichte des Universums in 100 Sternen by Carl Hanser Verlag, Munich, in 2019." | Includes bibliographical references and index.
Identifiers: LCCN 2021015389 (print) | LCCN 2021015390 (ebook) | ISBN 9781615197361 | ISBN 9781615197378 (ebook)
Subjects: LCSH: Stars--Miscellanea. | Astronomy--Miscellanea. | Cosmology--Miscellanea.
Classification: LCC QB801 .F74513 2021 (print) | LCC QB801 (ebook) | DDC 523.8--dc23
LC record available at https://lccn.loc.gov/2021015389
LC ebook record available at https://lccn.loc.gov/2021015390

ISBN 978-1-61519-736-1
Ebook ISBN 978-1-61519-737-8

Cover design and illustration by Beth Bugler
Text design by Jack Dunnington
Author photograph by Franzi Schädel

Manufactured in the United States of America

First printing August 2021
10 9 8 7 6 5 4 3 2 1

Contents

The Story of the Universe in 100 Stars

ARE 100 STARS ENOUGH to tell the story of the entire universe? No: The cosmos is far larger than we can possibly imagine and contains an equally unimaginable number of stars. Their true number is the subject of one of the hundred stories in this book, which, taken together, tell just one of many possible stories about the universe.

This book is not a mere inventory of the cosmos. Of course, you'll find out everything about the stars, galaxies, planets and all the other celestial bodies and phenomena you may encounter in the universe. You'll meet stars that tell of galactic collisions and reveal how black holes work; stars orbited by planets stranger than anything science fiction has to offer. Some stars allow us a glimpse of the beginning of the universe, others reveal what its future holds.

But a history of the universe is always also a story of mankind. Ever since we appeared on the scene, the universe has exerted a never-ending fascination on us. The stars have influenced our culture and our thinking and have made us what we are today— which is why the scientists who have expanded our knowledge of the universe matter, too. The stars in this book tell stories about famous people such as Isaac Newton and Albert Einstein, as well as about people you may not have heard of before: Dorrit Hoffleit, who first counted them; Henrietta Swan Leavitt, thanks to whom we know how big the universe is; Amina Helmi and her research into galactic fossils; Cecilia Payne-Gaposchkin, who discovered what stars are made of; Georg von Peuerbach, who paved the way for the heliocentric world view; and James Bradley, who proved once and for all that Earth revolves around the Sun. They, and the many others who show up in this book, have enabled us not only to admire the night sky, but also to understand it.

By the light of the stars, we can see how everything started 13.8 billion years ago, and how the Sun and the planet on which we live came into being. Their light has inspired us to invent myths and tell stories, spurred us to perform great technological feats and encouraged us to think deeply about what makes us who we are. Nowadays, it compels us to wonder whether we're alone in the universe and what our cosmic future might look like.

The hundred stars I have chosen for this book have little in common. Some are bright and have for thousands of years been part of the stories we tell ourselves about the heavens. Some shine so feebly that we can only discern them with the help of immense telescopes. Some have famous names, others merely bear catalog designations consisting of lots of digits and letters. There are large stars, small stars, nearby stars, remote stars. Some of the stories are about stars that haven't formed yet, others about ones that are long gone.

The stars are as variegated as the universe itself; each has its very own story to tell, and collectively they have shaped the history of the whole world. And that's exactly how this book works. You can open it to any chapter you like and immerse yourself in a partial story of the universe; each chapter has been conceived so that it can be read independently of the others. Or you can start at the beginning and read through to the end, and with each story delve deeper into the secrets of the universe.

The story of the universe is too complex to be encompassed by a single person in a single book. But the version told here with the help of the chosen hundred stars is one of the greatest stories ever told about the universe. It's the story of all those people who, over the course of millennia, have tried to understand the world in which they live—and the story of the fascinating discoveries made along the way.

Enjoy your journey through the cosmos.

HIKOBOSHI
The Cowherd and the Heavenly Weaver Girl

IT'S HARD TO MISS the brightest star in the Aquila constellation. Just sixteen light years from Earth, it is eleven times brighter than the Sun and the twelfth brightest star in our night sky. Its official name is Altair, which, like so many other stars' names, comes from Arabic.

During the eighth and ninth centuries, Arab astronomers expanded upon ancient Greek knowledge and published translations of the classical texts. When medieval European scholars translated these Arabic works in turn, they also adopted their designations for the stars. Thus *al-nesr al-ta'ir* ("the flying eagle") was the star we know as Altair today. Nearly every bright star in the sky has a name that stems from Arabic, including Ras Algethi, Algol, Dschubba, Fomalhaut, Mizar, Zubenelgenubi and many others. A handful bear Latin appellations, such as Polaris, Regulus and Capella. But even if Western culture very much rests on the foundations of Greco-Roman antiquity and its Arab reception, we mustn't forget that the sky has been studied at all times by all people.

Every culture thus has its own name for the stars and tells its own stories. In Japan, for example, Altair is known as Hikoboshi, and every year on July 7 a celebration is held in its honor. Or rather, in honor of Hikoboshi and Orihime—the cowherd and the weaver girl. Their story harks back to a 2,600-year-old Chinese folk tale: Orihime, daughter of Tentei, god of the sky, spends her days weaving garments for the gods. To provide his daughter with a little distraction, Tentei arranges her engagement to the cowherd Hikoboshi, but, like other young people in love, they consequently forget all about their work: The cows are running around unsupervised and the gods are waiting in vain for the cloth for their new robes. Tentei is forced to intervene and separate the lovers. He sends them into exile on opposite banks of the Amanogawa, the great heavenly river; yet there, too, their work remains undone, because Orihime and Hikoboshi are far too unhappy to concentrate on their duties. They are therefore granted permission to meet once a year, on the seventh day of the seventh month. However, when the time comes for their first visit, they find that there is no bridge across the heavenly river. Orihime cries so bitterly that a large flock of magpies takes pity on her and forms a bridge across the Amanogawa with its wings; they promise the couple that they will from then on do this for them every year on the seventh day of the seventh month—as long as there's no rain to make the heavenly river swell too much.

This sad love story and its happy ending are displayed in the sky to this day. Hikoboshi is, as I've said, the star Altair, and the heavenly weaver girl Orihime is represented by the bright star Vega; and just like in the tale, you can see the Milky Way—the heavenly river Amanogawa—stretch out between them. If you look very carefully, you can even make out the kindhearted magpies: The region of the Milky Way visible between Vega and Altair is partly filled with large interstellar dust clouds stretching in a dark band across the "heavenly river."

You can see Orihime and Hikoboshi high up in the sky in the summer, and particularly clearly at precisely the time that Japan celebrates the Tanabata. On that day, people commemorate the story of the cowherd and the weaver girl by erecting bamboo trees, to which they attach little pieces of paper containing their most ardent wishes.

The stars have inspired stories since long before we knew what they were all about. The sky is full of them, and we mustn't forget a single one, for, just as the stars tell us about the universe, the stories we tell about them reveal to us something of ourselves.

2MASS J18082002-5104378 B
Catching Sight of the Big Bang

THE STAR 2MASS J18082002-5104378 B is much more exciting than its rather unwieldy designation suggests. It's a small red dwarf just under 2,000 light-years from Earth and allows us a brief glimpse of the beginning of the universe.

When the universe emerged 13.8 billion years ago, there was only nothingness. Or not quite: Everything was potentially there, but not in its current form. Today, the universe is full of complicated things made from matter: large, hot spheres of gas orbited by smaller, cooler spheres, on which in turn (in one instance at least) even smaller beings have made themselves comfortable—beings that can, on rare occasions, also be spherically symmetrical. However, at the beginning of the universe there was no matter, and none of the many different kinds of atoms of which it's made.

Although we can't make any definitive statements about the moment of the Big Bang itself, we do have a fairly precise conception of the time immediately after. In the beginning it was enormously hot, and all that existed was energy and elementary particles. These had yet to arrange themselves into atoms as we know them today, which wasn't an easy thing to do: An atom is composed of one or more shells and an atomic nucleus consisting of positively charged protons and uncharged neutrons, which in turn consist of quarks. To the best of our knowledge, quarks are elementary particles—i.e., subatomic particles that cannot be divided into further particles.

The number of protons in an atomic nucleus determines which chemical element we're dealing with: One proton forms the nucleus of a hydrogen atom, helium needs two, lithium three, and so on. On the other hand, the number of neutrons in an atomic nucleus can vary and doesn't change an element's fundamental chemical qualities. For a complete atom you also need a shell around the nucleus, which contains the elementary and negatively charged electrons.

For atoms to be stable they require certain external conditions, which have only appeared in our universe over time. When the universe was in its infancy, it was still much too hot, and the enormous temperatures caused the quarks and electrons to move far too quickly for stable atoms to form. Fortunately, though, space cooled down very rapidly, and after just a hundredth of a second the newborn universe's temperature had decreased to a pleasant 18 billion degrees Fahrenheit (10 billion degrees Celsius)—enough to enable the quarks to cluster into protons and neutrons. A little while later, they in turn were able to form the first atomic nuclei, and thus the first chemical elements.

Hydrogen is a very straightforward element; nothing else has to happen for it to form, because a single proton already makes up the nucleus of a complete hydrogen atom. For helium, however, two protons and two neutrons have to find each other in the particle chaos of the young universe, and then connect. There's just one problem: Unlike a proton, a neutron flying solo through the universe is unstable, and destroyed by radioactive forces within a few minutes. After the Big Bang, then, the protons and neutrons only had a few minutes to find each other and fuse into nuclei, which is too little time for a complex nucleus to form. In the early universe there was thus an abundance of hydrogen nuclei (approximately 75 percent), somewhat less helium (approximately 25 percent), and here and there individual lithium and beryllium nuclei in vanishingly small numbers.

At least, this is what current cosmological theories about the Big Bang tell us about the ratio of chemical elements. We can only verify this through observation—for example by examining very old stars: The ones that were formed first in the universe can obviously only consist of the elements around at the time (i.e., hydrogen and helium) and in the aforementioned ratio. All the other chemical elements were created only later by the nuclear fusion that occurs in stars, so the older the star the more closely its composition has to match the ratio of masses present after the Big Bang.

And that is precisely what we have been able to observe: 2MASS J18082002-5104378 B is one of the oldest known stars—perhaps even the oldest. It was born just a few hundred million years after the Big Bang, and as predicted consists almost exclusively of hydrogen and helium, in exactly the expected ratio.

It's hard to believe that we can actually make concrete statements about the beginning of the universe, given how unbelievably long ago it all happened. But thanks to stars like this one such statements aren't science fiction—they can help us to confirm our theories and allow us to catch a brief glimpse of the moment when our cosmos began.

34 TAURI

The Planet That Was Once a Star

THE STAR 34 TAURI doesn't exist. Nevertheless, when the British astronomer John Flamsteed observed it in the sky he might have won great renown, if only he'd realized what it was he was looking at.

At the beginning of the eighteenth century, the astronomer royal was working on a large sky atlas at the observatory in Greenwich, which required him to systematically scour the sky and enter all the stars and their positions into a catalog. Theoretically, it was the ideal starting point for making new discoveries. He named one of those stars 34 Tauri—only it wasn't a star, but the planet Uranus. No one yet knew of its existence at the time. Uranus revolves around the Sun far beyond Saturn's path; it's as good as invisible to the naked eye, and even through a telescope barely distinguishable from a star. In contrast to real stars, however, it noticeably changes position over the course of several days.

The fact that the true nature of this spot of light eluded Flamsteed during his first surveys is understandable. They were separated by an interval of several years, and it's easy to see how he may not have worked out that it was always the same object, only in a different place in the sky. Moreover, his telescope wasn't quite good enough to show the planet as a small disk. All he saw was a dot, which looked just like all the other dots in the sky.

However, when Uranus came into his telescope's sights three times in the space of a single week in March 1715, he really should have noticed that there was something moving there, against the backdrop of stars. Yet he didn't—and because he failed to analyze his data closely enough, he was deprived of becoming the first person to discover the new planet. He shares this tragic fate with a few other near-discoverers of Uranus, including the German astronomer Tobias Mayer, who also registered the planet but didn't recognize it for what it was.

Nobody did, until sixty-six years later, when the British astronomer William Herschel spent March 13, 1781, in his garden monitoring the area around the Taurus constellation and noticed a spot of light that didn't quite belong there. Herschel built his own telescopes, and he built them better than anyone else, which is why he quickly saw that he wasn't dealing with a star. At first, he thought that the object was a comet, but his colleagues' calculations in due course confirmed that it was actually a planet, orbiting the Sun at nineteen times Earth's distance.

This was a sensational discovery. Until then, the only known planets were the six that are visible to the naked eye: Mercury, Venus, Earth, Mars, Jupiter and Saturn. It hadn't occurred to anybody that there might be more such large heavenly bodies revolving around the Sun. At a single stroke, Herschel's discovery of Uranus doubled the size of the known solar system and opened our eyes to the many things astronomy still held in store for us.

Out there were numerous new worlds waiting to be discovered, and Herschel's planet was just the start. Since then we have discovered asteroids, found Neptune and Pluto, and eventually even identified planets belonging to stars outside the solar system. And if John Flamsteed had been just a tiny bit more meticulous in his work, we could have begun our voyage of discovery sixty-six years earlier.

ALCYONE

Georg von Peuerbach and the Start of a Revolution

NOWADAYS, WE TEND TO TRACE the switch from a geocentric to a heliocentric world view back to the first decades of the seventeenth century, when Galileo Galilei's and Johannes Kepler's new findings made it clear that Earth isn't at the center of the universe; it moves around the Sun. But every revolution has a past.

Aristarchus of Samos already suspected in the third century BCE that Earth doesn't repose idly at the center of the cosmos. Our history, however, begins on September 3, 1457, when Georg von Peuerbach and his student Regiomontanus were standing together in the town of Melk in Lower Austria regarding the star Alcyone. There was a reason why the two astronomers' attention wasn't focused exclusively on the lunar eclipse that happened to be taking place at that moment: They weren't merely interested in watching Earth's shadow darken the full moon, they wanted to find out exactly *when* it would happen. Yet fifteenth-century clocks weren't the most precise instruments, so they resorted to the big clock in the sky. To do that, you need to know a thing or two about stars.

Georg von Peuerbach took up his university studies rather late, at the age of twenty-three, but was already lecturing at Italian universities three years later. He eventually became the first dedicated astronomy professor at the University of Vienna, where he busied himself with the creation of astronomical tables: This is what they called those long tables consisting of formulas and digits that they used to calculate the positions of the Sun, the Moon and the planets in the sky.

To verify the data in his tables he needed to take concrete measurements, which in turn required the occurrence of clearly definable and predictable events in the sky—for example the occultation of a planet by the Moon, or indeed a lunar eclipse, like the one due on September 3, 1457. To check that the formulas were correct, you had to know the precise time at which that event took place.

Georg von Peuerbach took advantage of the fact that Earth rotates on its axis once a day; or, as they still thought in those days, he used the fact that the firmament revolves around Earth once a day, while Earth sits motionless at the center of the universe. Regardless of how you look at it, though, in the course of the night you can see the stars climb higher and higher in the sky, and then, once they have reached the highest point, start to descend again. A star does this at a constant speed, equal to the speed at which Earth rotates on its axis. When you measure the height of a star above the horizon, you can therefore simultaneously determine the passage of time—and that's exactly why Georg von Peuerbach and Regiomontanus were watching Alcyone. This bright star belongs to the distinctive Pleiades star cluster, which you can see clearly even without a telescope. This was rather useful, of course, because the two researchers didn't have an optical instrument like that at their disposal; it was only invented 150 years later.

Even without one, though, Peuerbach was able to ascertain that the predictions made by the classical tables were not as accurate as they could have been. He began making corrections to the astronomical tables, developed mathematical methods for their improvement and built instruments with which you could determine the positions of the stars more accurately than ever before. His early death in 1461 meant that his work was left unfinished; it was later completed by Regiomontanus, who was really called Johannes Müller (he owed his nickname to the Latin translation of his birthplace, Königsberg, meaning "king's mountain"). He continued his mentor's work, producing even more precise calculations and developing even better mathematical techniques. He, too, died young—he was only forty when he died in Rome in 1476.

Three years before Regiomontanus's death, in the far north a boy named Niclas Koppernigk was born. He naturally knew nothing about Regiomontanus or Peuerbach yet, but when he grew older and began to preoccupy himself with astronomy, he based his work on their theories and on their data relating to the movements of planets. Today, Koppernigk is better known as Nicolaus Copernicus, the man who demonstrated that the Sun, and not Earth, is at the center of things and is orbited by the planets. This fact only became public in 1543, the year of his death, and it was another half a century before Galileo Galilei, Johannes Kepler and Isaac Newton established that a geocentric world view does not, in fact, correspond to reality. The "Copernican Revolution" thus happened long after Copernicus's time, and had started long before it.

FREISTETTER'S STAR

Can You Buy a Star's Name?

THERE IS NO "FREISTETTER'S STAR." Not a single one of the countless stars in the sky is named after me, and I'm quite sure that this won't change. I don't mind, though. I enjoy finding out about celestial bodies much more than the idea of having my name immortalized in the sky. Which, by the way, is not as easy as various online companies would like you to think.

These are the companies that offer you the opportunity to "Name a Star—Includes Registry and Certificate." For a small sum, you can pick any one of the many stars in the sky and call it whatever you like. You can name it after yourself, or after someone else, as a present. Once you've paid the required amount—and the brighter the star, the higher the amount—the name is entered into a "globally recognized star name register" or an "international star register." You receive an official-looking certificate and can then sit back and rejoice in the knowledge that a piece of the universe now bears a very special name.

Most of the rejoicing, though, is done by the companies that sell these certificates, for they're offering a product that doesn't really exist. The stars don't belong to anyone, and therefore nobody has the right to sell their names. Or, to look at it another way: Each one of us has the right to name them.

What we think of as the "official" names of stars are merely designations agreed upon by scientists, who consider them binding in their field. Aside from a few hundred stars that received their Greek, Arabic, and Latin epithets in antiquity or the Middle Ages, most only have a catalog designation consisting of digits and letters. For the scientific community, it is far more important and practical to have an orderly, unified nomenclature for the stars than to lend them poetic names.

Only in 2016 did the International Astronomical Union (IAU)—the global association for the promotion of astronomical research and collaboration—set up a working group for the naming of stars. The group's task is to standardize their names and, if necessary, assign new ones. So far, the official catalog has 336 entries, and only 6 are named after people: There are a Cervantes and a Copernicus, named after the Spanish author and the famous astronomer; Barnard's Star, after the astronomer Edward Emerson Barnard, who discovered that particular star's immense speed in the early twentieth century; Cor Caroli ("Charles's heart"), the brightest star in the Canes Venatici ("hunting dogs") constellation, after King Charles II; and then there are Sualocin and Rotanev, in the Dolphin constellation, named after the Italian astronomer Nicolaus Venator—they are his first and last names written backward, and have been called this since 1814, though nobody quite knows why.

These names were given to those six stars a long time ago and had became so commonplace that the IAU simply carried them over into their official catalog. There are a few other stars whose unofficial nicknames were inspired by real people, specifically the astronomers who discovered them: Tabby's Star, for instance (named after the astronomer Tabetha Boyajian), refers to the object with the catalog designation KIC 8462852, which created much excitement in 2015 when astronomers observed strange variations in its luminosity, which some attributed to alien activity. Once these nicknames have been used for a sufficiently long time by a sufficient number of members of the scientific community, it's possible that they, too, will be recognized by the IAU. But the commercial purveyors' "star registers" have nothing to do with the naming of stars. All you get for your money is an entry in some company database or other—which is neither binding nor unequivocal, because nobody can stop these firms from selling several names for the same star. Instead of spending money on the non-binding baptism of a star, therefore, you could do what people have done forever: Enjoy the view of the night sky and make up your own stories and names for the stars and constellations. Nobody can stop us from doing that. The sky belongs to us all.

HR0001

Mrs. Hoffleit Counts the Stars

"DO YOU KNOW how many stars there are?" asks a well-known children's song written in 1837 by the Protestant pastor Wilhelm Hey from the village of Leina in Thuringia. Hey immediately provides the answer—or at least, something like an answer: "The Lord God has counted them," the lyrics continue, but sadly the Creator doesn't reveal the result of his cosmic inventorying.

Where theology is at a loss, astronomy can help. However, astronomers do a lot more than just count the stars in the sky: They want to understand them in all their particulars. Yet to do that, they first have to catalog them. That's why any astronomical endeavor begins with a catalog listing as many characteristics as possible about as many stars as possible. For example, if you want to work out a star's mass or determine its age, you first need to know where it is, how bright it is and how quickly it moves. Catalogs seem boring, but they are the foundation upon which our knowledge of the universe rests.

On April 25, 2018, this foundation was significantly extended when the space telescope Gaia published the Gaia DR2 catalog, which lists no fewer than 1,692,919,135 stars. An impressive number, and a distinct improvement on the "mere" 2.5 million stars previously recorded in what was then the most comprehensive star catalog (Tycho-2). But our Milky Way alone consists of a few hundred billion stars, which means that the enormous Gaia catalog contains only around 1 percent of all the stars out there—even less, when we consider all the other galaxies in the universe. There are up to a quadrillion of these star systems in the visible cosmos, and each of them, too, consists of hundreds of billions of stars.

This means that there are a total of a few hundred septillion stars in the sky—at least in theory, for in practice we can't see most of them. Our telescopes are too weak and small to make them out. The most we can aspire to for the foreseeable future is to deepen our knowledge of the stars in our own galaxy—and even then it's highly unlikely that we'll one day have listed and counted all the billions of stars in the Milky Way.

For now, then, let's stay with the stars we can see without a telescope. Our eyes may be weak, but on a clear night they're eminently capable of observing an impressive firmament—as long as the artificial illumination from our cities doesn't outshine those natural sources of light. In a typical city, "the Lord God" would be quickly done with the counting: In densely populated and highly illuminated areas you can see no more than three dozen stars.

A glance at the Yale Catalog of Bright Stars shows us what things might look like under ideal conditions. It was compiled by the American astronomer Dorrit Hoffleit in 1956. In it, she listed all the stars that can theoretically be seen with the naked eye. Her extensive catalog begins with the star HR0001, which is about 530 light-years from Earth and glows so weakly that you can only see it if your eyes are very good indeed. All in all, the catalog comprises 9,095 stars, and all the relevant data known about them at the time.

Scientifically speaking, the correct answer to the question "Do you know how many stars there are?" is no. Nobody knows. Lots and lots, that much is certain. But only 9,095 of them can be seen with the naked eye—and those weren't counted by the Lord God, either, but by Dorrit Hoffleit of Yale University.

VEGA

Underrated Dust

DUST NEEDS AN IMAGE OVERHAUL: washing machines, dusters, vacuum cleaners—we have developed an entire industry here on Earth to rid our homes, our cars and our clothes of it. Astronomers, however, love dust. When it's far away out there in space, that is; down here they quite like to keep their instruments clean.

Cosmic dust is a unique source of information, as the astronomer Hartmut Aumann and his colleagues convincingly proved in 1984, when they used IRAS, the Infrared Astronomical Satellite, to examine the star Vega, the brightest star in the Lyra constellation and the fifth-brightest star in our night sky. Vega has been the focus of attention for astronomers for thousands of years and was on the list of stars to be inspected with the new space telescope.

Infrared radiation—the part of light invisible to the naked eye that is located beyond the red end of the color spectrum—can penetrate Earth's atmosphere only with considerable difficulty, because it is blocked out by the water in our atmosphere. To see it we therefore have to travel deeper into space, where nothing can interfere with our view. There is a lot to observe out there that we cannot apprehend with our eyes alone, and so IRAS became the first large space telescope designed to examine the sky with infrared eyes.

Stars don't radiate only visible light; they shine in every color, including those that are invisible to us without optical aids. Objects such as comets and asteroids, too, are warmed by the Sun, and reflect this heat in the form of infrared radiation; and then there's the dust: the cosmic small stuff out of which new stars and other heavenly bodies are formed. It's everywhere in the universe—between the planets, between the stars, even between the galaxies.

When the first data from IRAS arrived on Earth, astronomers were a little vexed, because some stars looked brighter than they should in the infrared light. The particular color of light emitted by a star radiating a given amount of energy is dictated by clearly defined laws. We can calculate and predict with considerable precision how much red, blue, green or indeed infrared light should be emitted by a particular star of a particular mass and temperature. Yet some stars evidently refused to comply with these predictions, especially Vega. There, they observed a phenomenon called "infrared excess," which simply means "too much infrared radiation." But Hartmut Aumann and his colleagues quickly realized that it was caused by dust. Of course, we can't see those tiny dust particles directly, but when they linger near a star they are heated by its rays and then reflect that heat in the form of infrared radiation. The whole thing then makes the star appear to be shining more intensely infrared than it should.

This meant that Vega was surrounded by a large disk of dust. But dust doesn't just come out of nowhere (although it seems to, in some homes). If Vega was covered in it, it had to come from somewhere—what's more, this sinister source had to be continually producing fresh dust, because when dust is left to its own devices it doesn't take long before it vanishes into space. All it takes to shift it is the light it meets.

So where there's dust, there have to be objects to create the dust: colliding asteroids, for example. This means that Vega can't merely be shrouded in dust; there has to be something else in its vicinity. Vega is in fact surrounded by asteroid belts, which shows that the same processes have taken place here as occurred in our solar system 4.5 billion years ago, which resulted in the formation first of asteroids and then of planets.

Until 1984, nobody knew for sure whether planets could form near stars other than the Sun; the dust surrounding Vega was the first confirmation we had that what happened in our corner of the universe also happens elsewhere, and that the search for the planets of other stars isn't futile.

RASALHAGUE
Confounded Astrologers

RASALHAGUE (ARABIC FOR "SERPENT'S HEAD") is the brightest star in the Ophiuchus ("serpent-bearer") constellation that regularly causes much excitement in the tabloid press and among astrology buffs. They claim that NASA has reclassified the star signs, and that there is now a thirteenth: DON'T FREAK OUT, BUT YOUR STAR SIGN MAY HAVE CHANGED, reads one headline from September 2016; DAILY HOROSCOPE BOMB-SHELL: YOUR ZODIAC IS WRONG SAYS NASA—HERE IS YOUR REAL STAR SIGN, reads another, from October 2018. According to these reports, anyone born between November 30 and December 18 is no longer a Sagittarius, but a Serpent-Bearer.

The reason for these regular bouts of media frenzy—aside from the tabloids' thirst for clickbait—is that many people are stumped by the difference between constellations and star signs as much as by the difference between astronomy and astrology.

We haven't always distinguished between them. In the old days, we used to watch the skies not only to study the movements and characteristics of all those dots of light we saw there, but also because we believed that heavenly bodies had mythological and religious significance. Comets, for instance, were for centuries thought to be portents of doom, or celestial accompaniments to momentous events (such as the birth of Christ). People thought that whatever happened in the sky had a material effect on human life, and that if we studied and understood the stars and planets well enough we'd be able to extract from them crucial information about the future.

It was only in the seventeenth century that the science of astronomy as we know it today began to diverge from the old superstition of astrology. These days, all those mythological figures and stories that people used to project onto the heavens in the form of constellations are only of historical relevance. Astronomy may still recognize the constellations, but they no longer have anything much to do with astrology's star signs.

The twelve "signs of the zodiac," to give them their proper name, which are drafted into the service of newspaper horoscope columns to this day, are constellations situated in a very specific place in the sky. They are ranged along the ecliptic—i.e., lined up along the Sun's apparent annual route through the sky. The ecliptic is simply Earth's orbit around the Sun, projected against the sky; and the rest of the solar planets also move in, or close to, this plane. This is the reason why the twelve signs of the zodiac used to play such an important role in ancient surveys of the skies: People were able to observe when and how the planets passed through the individual constellations and draw astrological conclusions from this.

Just like the constellations of Scorpius, Sagittarius, Aries, etc., the Ophiuchus constellation also stems from antiquity, and it, too, is traversed by the ecliptic. But for some reason it was never included among the official signs of the zodiac—perhaps because they thought twelve a more auspicious number than thirteen.

For a long time, there was no binding regulation governing the constellations. There was no written record of which constellations existed, or a list of the stars in each constellation. It took until 1928 for scientists to put things right, when the International Astronomical Union divided the sky into eighty-eight discrete areas and thus created the eighty-eight constellations still officially recognized today. The twelve constellations along the ecliptic whose names correspond to the twelve star signs have remained; and so has Rasalhague, in the Ophiuchus constellation.

Which isn't to say that Ophiuchus is an astrological star sign, too. Astrology frequently ignores the insights gained by astronomy, and Ophiuchus is no exception. While the constellations occupy areas of different sizes in the sky, the astrologers' star signs are all the same size. This results in substantial discrepancies: for example, someone whose star sign is Aquarius will assume that, astrologically speaking, the Sun was in Aquarius's part of the sky at the time of their birth; from an astronomical perspective, however, the Sun was in Pisces.

Astrology isn't a science, and star signs have nothing to do with the official constellations, or with the Sun's actual position in the sky. Nevertheless, the star signs have their own part of the cosmic story to tell—they are testament to the human urge to put our lives in some kind of celestial context. But there's no need to worry that you might have been living under the wrong star sign, though you might well ask why we still place such value on an ancient superstition.

TXS 0506+056

Ice Cube Astronomy

AT EARTH'S SOUTH POLE there is what must be the world's strangest telescope. Or, to be more precise, it's more than half a mile (about a kilometer) underneath the South Pole, frozen into the Antarctic ice. It's searching for a very special type of light down there. What it's trying to find is neutrinos—something that requires a considerable effort.

A neutrino is an elementary particle. It has almost no mass, and hardly interacts with other matter. Around a hundred million of them are racing unnoticed across every tenth of a square inch (or across each square centimeter) of the surface of our bodies at any given second. The entire Earth, in fact, is exposed to a steady stream of neutrinos, but without being in any way influenced by it. As far as these elementary particles are concerned, Earth isn't merely transparent—it's as if it doesn't exist at all.

Neutrinos are created by nuclear reactions, and in especially vast numbers by the nuclear fusion that occurs inside stars: The Sun produces not only the light it radiates into space, but also a continuous flow of neutrinos, which are much harder to see. The numerous neutrinos that continuously hit our planet only rarely interact with normal matter, but when they do, these collisions form new particles, which then briefly emit energy in the form of a bright light. However, for us to witness an event like that a number of other factors have to coincide: There has to be a lot of matter around for the neutrinos to encounter something in the first place; and we need special detectors, or rather very many highly sensitive optical sensors, capable of capturing these brief flashes of light.

Our telescope at the South Pole is just such a neutrino detector, and one of the reasons that it's buried so deep in the ice is that the pressure from the overlying layers squeezes even the tiniest air bubbles out of the ice, making it extremely clear and transparent. This is the only way for the optical modules to be able to spot the flashes of light. Distributed across a volume of 0.24 cubic miles (1 cubic kilometer), they have sunk 5,160 sensors into the ice; it's almost as if a gigantic subterranean block of ice has been converted into a neutrino detector—which is why the project is called IceCube.

The data collected by the sensors is transmitted via cables to the surface, where it is then evaluated. From the intensity of the flash we can calculate how much energy the neutrino in question is carrying; and the greater the energy, the more powerful the event during which the neutrino was produced must have been. In addition, the IceCube allows us to work out from which direction the neutrino came. Most of the neutrinos detected there originate from the Sun: It is by far the most prolific source of these particles in our neighborhood, and we were able to observe the neutrinos it produces long before IceCube. But it's only thanks to IceCube that we now fully understand how the Sun creates energy, and the specific kind of nuclear reaction that takes place inside it.

Naturally, we would also like to observe neutrinos produced by other celestial bodies in our universe, and we did this in 1987, when a supernova occurred 170,000 light years away. We detected a total of twenty-five neutrinos at the time—which doesn't sound like many, but constitutes a veritable particle storm in the field of neutrino astronomy.

We catch sight of these elusive particles so rarely that we can give them individual names. In 2013, IceCube observed two neutrinos of such high energy that they couldn't possibly have originated in the Sun; they must have come from the outer reaches of the universe. The scientists were unable to determine their origin but called them Ernie and Bert.

On September 22, 2017, they found another high-energy neutrino, but this time, instead of naming it after a character from *Sesame Street*, they gave it the sober designation IceCube-170922A. Immediately after IceCube detected it, ordinary telescopes all over the world set out to discover its source—and found it: Exactly in the direction it came from, you can see TXS 0506+056, a "blazar," shining in the sky. Blazars are the extremely active centers of remote galaxies, which contain supermassive black holes; when matter vanishes into them, it sends out a tremendous blast of energy, creating neutrinos in the process—and it looks like IceCube identified just such a neutrino. The neutrinos' extreme reluctance to react with the rest of matter makes them extremely valuable to us. They are formed in the stars, in the centers of other galaxies and in the vicinity of black holes. Unlike light, they are well-nigh unstoppable, and can therefore carry information about their origin unhindered across the whole universe.

The subterranean ice cube located at the South Pole is the best neutrino telescope in the world. It has enabled us to observe the skies in an entirely new way—though if we want to see the cosmos *really* well in the light of the neutrinos, we'll need a bigger detector.

π¹ GRUIS

A Simmering Giant

THERE'S A GIANT STAR bubbling away in the Grus constellation, 530 light-years from Earth. Its name is π¹ Gruis, and if it was in our solar system it would have swallowed Earth long ago. It has almost reached the end of its life, but not without first showing us something we've never observed before in this form.

The only star we can examine in proper detail is the Sun. It may look like a simple white disk, but when you consider it a bit more closely through a telescope you'll recognize it for what it really is: an enormous, hot, simmering mass of gas. What you see happening on its surface is the same thing that happens when you boil a saucepan full of water. Think of the Sun's center as the stove, where hydrogen nuclei are fusing into helium nuclei; that's where the Sun produces all its energy. First, this energy spreads out in the form of high-energy light particles, in a "radiation zone" extending to around 300,000 miles (500,000 kilometers); this far from the center, the temperature has already decreased from about 27 million degrees to 2.7 million degrees Fahrenheit (15 million degrees to 1.5 million degrees Celsius). The energy is then transported by the heat across the remaining 125,000 miles (200,000 kilometers) to the Sun's surface. The radiation heats up the gas in the Sun, which starts to rise. As it does so it cools down, until it reaches the Sun's surface at a temperature of just 9,925 degrees Fahrenheit (5,500 degrees Celsius). Now that it has cooled down, the gas sinks back toward the center until it once again grows hotter, and the process is repeated.

This cycle is called "convection," and is exactly what happens when we heat water in a saucepan. The Sun's individual convection cells—that is, the zones in which its material rises to the surface—measure about 600 miles (1,000 kilometers) across and are called "granules." When you look at a time-lapse video of the surface of the Sun, you can watch it literally bubbling away.

What is comparatively straightforward in the case of the Sun proves to be almost impossible in the case of other stars. They are too far away for us to perceive them as anything other than spots of light, even through large telescopes. Their surface structures and the convection processes that occur there usually remain invisible to us. But in 2018, we managed to observe them on π^1 Gruis. Not only is it a huge star, it's also a thousand times brighter than the Sun—so luminous that, when the European Southern Observatory in Chile combined the data from four different telescopes to create a virtual telescope of much greater dimensions, Claudia Paladini from the University of Brussels and her colleagues were able to "resolve" the star's surface and make its convection cells visible.

The cells on π^1 Gruis are as colossal as the star itself: The granules measure close to 75 million miles (120 million kilometers) across, which is almost the same as the distance between Earth and the Sun. The reason for this is the star's low density: It may have one and a half times the Sun's mass, but it's much bigger than the Sun—and therefore less dense. This is also why the gravitational force is weaker on its surface than on the Sun's, which results in larger granules.

This behavior had already been predicted by theoretical star models but was definitively observed for the first time on the surface of π^1 Gruis. Some of it was down to sheer luck: The phase in which the giant star currently finds itself lasts just tens of thousands of years, a hundred thousand at most. π^1 Gruis has already shed a substantial part of its atmosphere into outer space; in a few tens of thousands of years it will have disappeared completely, leaving behind only a small, dead star-remnant. We caught it just in the nick of time to watch it bubbling away.

11

B CASSIOPEIAE

A Dogma Blows Up

THE STAR B CASSIOPEIAE no longer exists. Even while it was alive, nobody knew of its existence. We were only able to see it when it disappeared—and took an old dogma with it in the process.

It's actually wrong to describe B Cassiopeiae as a star. However, the Danish astronomer Tycho Brahe didn't know that when, on November 11, 1572, he saw a brightly shining object he'd never observed before in the Cassiopeia constellation. But there it was, almost as bright as Venus, and impossible to miss. It was a "new star," and he wrote an eponymous book about it that made him one of the most renowned astronomers of his time. In *De Stella Nova*, he proved that the bright object wasn't—as many of his contemporaries thought—merely a mysterious light phenomenon in Earth's atmosphere but had to be located somewhere among the stars.

After all, if the light were somewhere in our atmosphere, close to Earth, it would have to move against the backdrop of stars in the sky as Earth rotates around its axis once a day. (Or rather, according to the generally accepted view at the time, the light would have to be standing still along with Earth, while the rest of the universe revolves around it once a day.) But this wasn't the case: The light was moving in unison with the stars. Tycho Brahe was therefore convinced that it was a new star.

Brahe was not the first, or the only one, to see it, but no one else at the time examined it as closely as he did. With his precise observations, he was able to refute the millennia-old dogma of the unchanging celestial realm. The Greek scholar Aristotle once proclaimed that the heavens had to be perfect, and forever immutable; change could exist only on Earth. The Church later concurred with him—until Brahe's observations showed that even in the heavens things can change.

However, he was wrong to describe the star as "new." We know now that an event like this indicates an old star, blazing its final salute to the universe. Brahe had observed what we now call a "type Ia supernova." This requires a white dwarf, which is a star that has already shut down its nuclear reactor: It has used up all the fuel that powered its nuclear fusion process, cast off the outer layers of its atmosphere, and now no longer does much of anything except exist and cool down. But when this white dwarf is part of a binary star system, and when the two stars are very close to each other, the gravitational force of attraction that operates between the two celestial bodies can cause material from the other star to land on the white dwarf.

And then the white dwarf's nuclear fusion process starts up once again, and it again begins to shine. However, the star—in effect already deceased—is no longer capable of regulating its temperature in the same way as a normal star. Within a short time, it becomes immensely hot, and at the end literally explodes. This explosion is the "supernova" we then witness in the sky, which Brahe took to be a new star. It takes a few weeks or months before the light emitted by the explosion goes out, and with it the "new star" vanishes once again from the sky. In this case, though, the trace it left changed our conception of the universe forever.

12

ACRUX

One Star, Too Many Names

ACRUX, BECRUX, GACRUX AND DECRUX. They might sound like characters from the comic series *Asterix* but are legitimate astronomical terminology. Acrux and Co. are the official names of stars belonging to a very famous constellation: Crux, the Southern Cross.

As the name suggests, you can see the Southern Cross particularly well in the southern hemisphere. It's small, but distinct—so distinct, in fact, that today the flags of Brazil, Australia, New Zealand, Papua New Guinea and Samoa carry its image.

The strange names of the stars that make up this constellation are in a certain regard the result of a lack of respect. When, in centuries past, explorers from Europe and North America set out on their voyages to the southern hemisphere, they actually "discovered" very little there, in the true sense of the word. People had lived there since time immemorial and given the things around them names, categorized them and incorporated them into their histories and their stories. Yet that didn't prevent the explorers from entirely ignoring the existing names of countries, mountains, rivers, animals, plants—and stars, too—and simply furnishing them with new ones.

As far as the systematic labeling of stars is concerned, astronomers have long favored the so-called Bayer system. It was devised by the German astronomer Johann Bayer, who applied it in his influential 1603 sky atlas *Uranometria*. He had organized the stars into constellations and then arranged them according to brightness; by this method, a star's designation is thus constructed from the Latin name for the constellation and a Greek letter: "Alpha" for the brightest star, "Beta" for the next brightest and so on.

The brightest star in the Centaurus constellation is therefore called Alpha Centauri, and the third-brightest star of the Cepheus constellation is Gamma Cephei. Continuing in this vein, then, the Southern Cross consists of the stars Alpha Crucis, Beta Crucis, Gamma Crucis and Delta Crucis. Or Acrux, Becrux, Gacrux and Decrux for short—forms that eventually became so well established that they were adopted as the official designations for those stars.

And what once becomes official remains official—even though we have now realized that this naming method not only shows a lack of imagination, but also reflects the insensitivity with which so many traditions have been ignored—the traditions of people and peoples who had observed and named these stars long before Europeans ever thought of sailing into the south.

In 2018, the International Astronomical Union tried to rectify the situation a little. The fifth-brightest star in the Southern Cross, which till then bore the Bayer designation Epsilon Crucis, managed to escape the somewhat ignominious fate of being called Epcrux. It had yet to be given its official name, so the IAU made the most of the opportunity and called it Ginan. In the stories of the Wardaman people of northern Australia, a *ginan* is a traditional bag filled with stories and songs and myths about the creation of the world. And now it can officially be found in the sky, too—as well as on the Australian flag, which includes Ginan alongside its four drably named fellow stars.

13

51 PEGASI

The Answer to a Thousand-Year-Old Question

"WHETHER THERE BE ONLY ONE WORLD or many worlds, is one of the most wondrous and noble of nature's questions. It is a question that the human spirit has to answer for itself," wrote the German bishop and scholar Albertus Magnus in the thirteenth century. The answer to this wondrous and noble question was only discovered more than seven hundred years later—thanks to the star 51 Pegasi.

The question of the existence of "other worlds" has preoccupied humanity for thousands of years. Even the pre-Socratic philosophers of ancient Greece wondered whether Earth was the only "world," or just one of many. The atomist Democritus, for instance, postulated a limitless number of worlds, some of which have more suns and moons than ours, others fewer. The discussion continued during the subsequent centuries, and preoccupied philosophers as well as theologians throughout the Middle Ages and into the early modern age. But regardless of whether they invoked God or nature, they were unable to reach a definitive conclusion.

One can argue the matter from a theological point of view, and say, for instance, that there is no mention of other worlds in the Bible, so they simply cannot exist. Or that, on the contrary, an all-powerful, omnipresent, infinite "God" can't help but create infinitely many worlds. Over the centuries, theologians, philosophers and natural scientists found innumerable arguments for and against the existence of other worlds. But argue, speculate and discuss was all we could do. As yet, we knew too little about the world.

We first had to realize that Earth is a planet that orbits the Sun, which is a star. Then we had to find out that all those spots of light in the sky are also stars, similar to the Sun but much farther away. We had to understand how stars and planets come into existence and acquire the right instruments to examine them.

That's why it took until the 1980s for scientists to start looking for "other worlds" in earnest—or, as we call them these days, extrasolar planets (i.e., planets that orbit stars other than the Sun). We knew that they could exist in principle. But because we didn't yet fully understand how and under what conditions planets are formed, we weren't sure how many of them there are. Was the Sun, with its eight planets, a cosmic exception? Nevertheless, scientists began their search, and in the 1980s and 1990s astronomical working groups from the US and Canada competed for the glory of having discovered the first exoplanet.

As it happened, the winners of the planet hunters' race proved to be two outsiders: Michel Mayor from the University of Geneva, and his PhD student Didier Queloz. In April 1994, they focused their telescope on the star 51 Pegasi, which is fifty light-years away from us in the Pegasus constellation. It's a little larger and older than the Sun, and under the right circumstances you can just about perceive it with the naked eye—but the planet discovered there by the two Swiss astronomers is of course invisible without the help of technology. Actually, you can't even see it *with* the help of technology: Mayor and Queloz discovered the celestial body only indirectly, when they noticed that the star was wobbling in a very particular way. Every planet exerts a gravitational force—if only a very small one—upon a star as it revolves around it, which makes the star move a little to and fro. Its wobble can be measured by analyzing the starlight, and from this you can deduce the planet's mass and orbital path.

However, when you announce that you have found the answer to a millennia-old question, you have to expect a few skeptical looks; and Mayor and Queloz's discovery was initially regarded with suspicion. However, independent observations have confirmed the existence of this controversial planet several times since: It's twice as large as Jupiter, only half as heavy, but just as real. We finally knew: Other worlds do exist. Other stars are also orbited by planets, and 51 Pegasi was the first star to corroborate our long-held hunch. That's why, in 2015, it was among the stars to receive a new name from the International Astronomical Union; it's now known—hardly surprisingly, given its discoverers' nationality—as Helvetios.

14

61 CYGNI
Killer of the Crystal Spheres

DURING THE FIRST HALF of the nineteenth century, 61 Cygni abruptly found itself at the heart of one of the most important scientific questions of all time. It's a binary star system in the Cygnus constellation, a touch smaller and a little older than the Sun, and in the eighteenth century they thought that it had to be located reasonably close to the Sun. Probably. For this was precisely the big unanswered question: How far away are the stars? Or, in other words: How big is the universe?

You can readily see the stars shining in the sky at night, but determining their distance from Earth isn't quite as easy. Greek scholars—including Anaximander in the sixth century BCE and, later, philosophers like Pythagoras and Aristotle—believed that the stars were attached to a kind of crystalline sphere that revolved around Earth, which was sitting immobile at the center. This heavenly sphere at the same time delineated the edge of the cosmos, which, according to this conception, was rather small. Observations made during the sixteenth century, however, demonstrated that such a crystal shell couldn't possibly exist, and that the distance to the stars had to be greater than previously thought.

Again and again, scholars like Giordano Bruno speculated that the lights in the sky suggested the existence of Sun-like objects, which only appeared to us as tiny dots because they were so far away. But nobody knew how to measure the distance to the stars, and thereby answer the question once and for all. That is, there was one way they knew of, but it didn't work.

In the seventeenth century, the heliocentric world view grew dominant. Nicolaus Copernicus, Galileo Galilei and Johannes Kepler made it eminently clear that Earth revolves around the Sun, not the other way around. But if this was true, we would be looking at the distant stars from very different positions at different times during the year—which would give them the appearance of being in motion.

You can very easily test the principle of the "stellar parallax" method yourself: Stretch out your thumb and the arm to which it belongs, and then look at the thumb, first only with your left eye and then only with your right. Because your eyes are set slightly apart from each other, each is looking at the thumb from a different point, and it seems like your thumb is leaping to and fro against its background.

This is what the stars ought to be doing, too, when we observe them at different times of the year—and thus from different positions in the solar system. The greater the apparent movement, the closer to us they must be, and the parallax effect should thus enable us to measure their distance.

But nothing was moving.

At the time, the obvious explanation for this was that Earth was fixed to the center of the cosmos after all. But a different hypothesis was quickly developed; the absence of any "movement" in the stars in the sky might simply be due to the fact that the stars are many times farther away than had been thought: The greater the distance, the less their apparent position would change.

The star system 61 Cygni was the perfect research subject for another go: It's bright, and thus easy to observe. However, for a while no one managed to obtain any reliable measurements from it. When the German astronomer Friedrich Wilhelm Bessel performed his first attempt at a parallax measurement in 1812, this, too, failed, but when he repeated the experiment with a better telescope in 1837 and then again in 1838, he finally succeeded: For the first time, a star's parallax had been conclusively observed, and its distance measured. According to Bessel, it was 9.25 light-years away—not at all far off the current most accurate measurement of 11.4 light-years.

Only twelve other stars are closer to us than 61 Cygni, which means that the great majority of stars are a lot farther away. By determining its distance, Bessel showed that the real universe has nothing whatsoever to do with the cozy crystal-sphere cosmos of antiquity. Bessel might not have been able to determine its actual size, but he'd taken the first step into its vast expanse.

BPS CS 22948-0093

A Cosmic Shortage of Lithium

THE STAR BPS CS 22948-0093 has a lithium problem. It contains too little of this chemical element—something that's true for other stars, too. It's an old star, situated nearly 7,000 light-years away in the outer reaches of our Milky Way. Stars like this one belong to an earlier generation than our comparatively young Sun. They came into being at a time when the universe was a simpler creature than nowadays—at least as far as its chemical composition is concerned.

The first atoms formed 13.8 billion years ago, immediately after the Big Bang. However, the existing elementary particles only had a few minutes to produce complex atomic nuclei: the right conditions didn't really persist long enough for anything other than hydrogen and helium—the simplest chemical elements—to emerge. The nuclei of all other elements are so complicated that they didn't emerge until much later, and in much smaller quantities, from the nuclear fusion occurring inside stars. With one exception: The atomic nucleus of lithium is just about simple enough to have allowed an extremely small amount of this element to form directly after the Big Bang. (A trace of beryllium was also created, but it plays no part in this story.)

At the beginning, then, all there was in the universe was a lot of hydrogen, a little less helium and a pinch of lithium, and the composition of very old stars should reflect this ratio, too—for they can inevitably only consist of whatever materials were around at the time. As far as the ratio of hydrogen to helium is concerned, the assumptions of the Big Bang theory do indeed closely match what we can observe in old stars. But lithium poses a problem: In stars like BPS CS 22948-0093, we have measured two to three times less of it than there should be.

One possible explanation for this is that we haven't properly understood the formation of elements after the Big Bang. Yet this is unlikely, since so much observational data has again and again backed up what we think we know about the early days of the universe. A more plausible explanation is that the lack of lithium has something to do with the processes that take place inside stars.

Lithium can be destroyed during the nuclear fusion that occurs in a star, even at comparatively low temperatures, so we can in any case expect there to be less of it around today than there might have been before. However, the unexpectedly small quantity of lithium may also be ascribed to the fact that old stars are much more effective than young ones in their destruction of it. If the lithium present in the cooler outer layers of a star is continually transported toward the core, and if the lithium deep down in a star's hottest layers also disappears over time, this might explain the general lack of lithium.

Assuming that this element is indeed redistributed inside a star by means of some process of diffusion, older stars ought to contain less lithium, because they'll have had more time to destroy it. Moreover, hotter stars, which develop more quickly, should also show less lithium than cooler stars of the same age, whose nuclear processes are proceeding more slowly. Corresponding observations suggest that this is precisely the case—but it isn't clear to us yet what is causing the diffusion itself.

We'll have to wait a bit longer for a definitive answer to the enigma of the cosmic lack of lithium; in the meantime, we're going to keep an eye on stars like BPS CS 22948-0093—for, who knows, we may still learn something from them about what really happened right after the Big Bang.

16

62 ORIONIS

Caroline Herschel Emerges from Her Brother's Shadow

THERE WAS A FLY in the ointment at 62 Orionis. There's nothing wrong with the star as such: It was discovered at the beginning of the eighteenth century by the British astronomer John Flamsteed during his large-scale survey of the skies, and he had earmarked it for inclusion in his catalog—where it ended up as the sixty-second item in the list of stars that form the Orion constellation. The problem with it, rather, was all those stars that had failed to make the catalog. And it is this mistake, as well as all the other mistakes and overlooked stars, that Caroline Herschel set out to correct.

At first, it didn't look like Caroline Lucretia Herschel, born in 1750 in Hanover, would embark upon a scientific career. She was the youngest child of a large family and received only a rudimentary education. Beyond that, she wasn't much more than her family's maid; she took care of all the housework, while her big brother William moved to England to pursue a career in music.

When Caroline was twenty-two years old, William invited her to join him in England. He told her that she'd be able to make a living as a singer—which she no doubt could have done, if she hadn't been too busy looking after her brother and managing his household. He in any case gave up the music thing quite quickly, and from then on dedicated himself to his second-favorite hobby: astronomy.

Once again, it fell to Caroline to support him in his work. She copied out scientific texts for him, recorded his observational data, made calculations and ground mirrors for telescopes, and drew great pleasure from her work. When her brother discovered Uranus in 1781 and attained worldwide fame overnight, the siblings were able to dedicate themselves exclusively to astronomy; and when the king appointed Herschel as his court astronomer, Caroline became his official assistant. She was paid a salary, and now also had her very own telescope at her disposal, with which she soon pursued her own projects. In August 1786, she discovered the first of a career total of eight comets, which was a considerable number in those days.

She combed the entire sky in her hunt for unusual objects and started correcting the data her predecessors had collected. At that time, Flamsteed's was the most famous and important star catalog—but it was also incomplete. Flamsteed had died before its publication, and his successors had evidently not been as assiduous in their handling of his data as they should have been. So Caroline began again at the beginning and compiled a new catalog, which now contained all the stars observed by Flamsteed that hadn't been included in his sky atlas. Her work was published in 1798 and started with the star 62 Orionis and the omitted stars in its vicinity. Once Caroline had sorted and added the missing observational data, she rectified the mistakes that had crept into Flamsteed's catalog during its creation. Once again, order reigned in astronomy; all the necessary data was now assembled in one place, and the search for new astronomical discoveries thus became much easier.

Contrary to what her biographers have claimed, Caroline Herschel wasn't the first female scientist to be paid for her work. Nevertheless, she was a true pioneer, and she continued her scientific work long after William's death in 1822. Yet she had to wait a while for her colleagues to allow her to step out of her brother's shadow: She was seventy-eight when she was awarded the Royal Astronomical Society's Gold Medal, became the society's first female honorary member at eighty-five, and was made an honorary member of the Royal Irish Academy at eighty-eight; it wasn't until she reached the ripe old age of ninety-six that she was finally honored in her native Germany, too, and awarded the Prussian Academy of Sciences' Gold Medal. But if Caroline proved one thing in her life, it was her stamina.

17

ANTARES

Fluff in the Superbubble

THE UNIVERSE IS basically quite empty. Leaving aside for the moment the stars strewn across it at great distances, nothing much happens in all that vast interstellar space. But if you know precisely what you're looking for, and where, you can still discover enough to attract the attention of astronomers. A few hydrogen atoms, for example—no more than a handful per a tiny fraction of a cubic inch, roughly a cubic centimeter of space—or one or two helium atoms. With a bit of luck, you'll even come across traces of other chemical elements. All these widely scattered particles are known as the "interstellar medium," and it's well worth taking a closer look at it, because those places where the interstellar medium is a little more or less dense than elsewhere hide clues to how the universe works. The denser, cloudlike expanses between the stars are the very regions where new stars can come into existence, and wherever the interstellar medium is much thinner than elsewhere, dying stars have had a metaphorical hand in the game.

Our solar system occupies an area with a radius of about 300 light-years known as the Local Bubble. Here, the interstellar medium is very thin indeed; you can find almost nothing but hydrogen and helium, and as good as no dust—"dust" being what we call those areas of the medium that consist not of individual atoms, but of more complex structures (such as carbon atoms that have bonded together to make graphite) as well as minute ice particles. They form everywhere that planets also form and are sometimes pushed off into space by stellar radiation.

This is the reason why there's hardly any dust left in the Local Bubble. Several large stars must have concluded their lives here during the past few million years; like other large stars, they would have met their end in big explosions that positively swept the dust out of their neighborhood, thus creating the Local Bubble. But in the middle of this bubble, a little bit of dust has remained—we call this the Local Interstellar Cloud, aka the Local Fluff. This titanic bit of fluff has a diameter of about thirty light-years, and it is stars like Antares that are responsible for its existence.

Antares is the brightest star in the Scorpius constellation, and easily recognizable from its reddish light. It's approaching the end of its life, and has ballooned into a mighty giant that, if you substituted it for the Sun, would fill the entire solar system as far as Mars's orbit and beyond. Even though its life as a star is nearly over, Antares is still relatively young. It was born only about 15 million years ago, but massive stars like this one are enormously hot and therefore use up their material much more quickly during nuclear fusion than small stars like the Sun.

Antares belongs to the so-called Scorpius–Centaurus Association, a group of a few thousand similarly young and bright stars. They are all part of the same star-forming region, and as young stars are wont to do, when they were born they whirled up all kinds of dust and gas and catapulted it out into space.

Our solar system is currently moving through this material and has done so for about 100,000 years. It will take another few tens of thousands of years before it leaves this cosmic bit of fluff behind and moves on to the somewhat cleaner regions of the Local Bubble. But this voyage through the fluff has no noticeable effect on us down here on Earth. It would take more interstellar matter than that to bother us.

18

HAIRY STARS

Portents of Death and Messengers from the Past

IN THE OLD DAYS, people would get extremely worried when they saw a "hairy star" in the sky. It was a sight they weren't used to. Instead of the usual, faithfully shining dots of stars and planets, here there was a wild, cloudlike thing dragging a long tail after it, looming over Earth. It would show up without warning, and then simply evaporate again. They called these objects "fiery besoms" and "faxed stars." Or "comets," from the Greek κόμη (*kóme*), referring to the hair on your head.

For a long time, nobody knew what these "stars" and their manes were all about. The Greek scholar Aristotle, for example, thought that they weren't celestial bodies at all, but atmospheric light phenomena. In the seventeenth century, the general consensus seemed to be that, whatever it was, a comet was not a Good Thing. A comet would be followed by no less than "much fever, malady, pestilence and death, hard times, destitution and great famine, great heat, droughts and infertility, war, robbery, fire-raising, murder, convulsion, envy, hate and strife, frost, chill, tempest, evil weather, thirsting, the ruin and death of great men, conflagration and earthquakes," declared the pastor and astronomer Johann Gottfried Taust in 1681, citing a well-known poem of the time.

From the point of view of the age, this anxiety was understandable. After all, the stars could clearly be seen sitting immobile in their assigned places in the firmament; and although the planets moved, they did so in a reasonably predictable way, even in those days. Only the comets evidently refused to conform to the rules, and their appearance and disappearance confounded any attempt at prediction or explanation.

It wasn't until Isaac Newton conceived his mathematical theory of gravity that it became possible to demonstrate that comets, too, orbit the Sun. Their orbits are often merely somewhat elongated, not the near-circles along which the planets move. Comets can also put a lot of distance between themselves and the Sun, which means that we only see them when they happen to be close to it.

Today, we know exactly what comets are: They're the "rubble" left over from the time when the planets of the solar system were formed, 4.5 billion years ago. When the dust and gas around the Sun clumped together and eventually created the planets, a little of all that material was left over: lumps of rock and ice, typically no larger than a couple of miles across. Comets are those chunks that contain a large amount of frozen material, and as they get closer to the Sun the ice heats up, turns into gas and streams into space; in the process, it rips dust from the comet's surface, which then envelops the comet. This large shell of dust, the "coma," reflects a large amount of sunlight, which then makes the small celestial body visible. The dust can also create the comet's familiar tail, which gives it its singular appearance in the sky.

These days, comets have largely lost their superstition-inspiring aura (although some people still think of them as omens of the end of the world, a sign from God or alien spaceships, as the case may be). Instead, we now recognize them as invaluable messengers from the past. They're the last survivors from an ancient past when the solar system was still young and devoid of planets. By studying them, we can find out much about how it all started back then. Comets are not harbingers of doom, but emissaries from our beginnings.

19

HD 142

Our Bright Astronomers Frequently Generate Killer Mnemonics

"CLASSIFICATION OF 1,688 SOUTHERN STARS by Means of Their Spectra": The title of this 1912 paper by the American astronomer Annie Jump Cannon and her colleague Edward Pickering doesn't sound particularly imposing. And when you start reading it, at first there seems to be nothing very earth-shattering about it. Its fifty pages contain hardly any words but are instead covered in tables showing the coordinates and characteristics of various stars. The first in the list is the star with the designation HD 142. The ninth column of the table gives it a letter and a number—G0—and all the other 1,687 stars have also each been assigned a letter and a number.

These designations may appear to tell us nothing, but they actually reveal a lot about the characteristics of the relevant star. "G0" is the so-called spectral class of the star HD 142, which describes the quality of the star's light with regard to the light spectrum. Stars emit light of every color into space; using suitable instruments, we can break this light down into its component parts—i.e., turn the star's blended light into a rainbow. If you look closely, you can perceive dark lines in the light spectrum, the "spectral lines," which are caused by the chemical elements that make up the star. When a star's light penetrates its gas layers, a portion of it is blocked out by the atoms in those layers. Each chemical element blocks out a different, entirely unique portion of the light—so the spectral lines are like a bar code, a definitive chemical fingerprint that tells us which elements a star consists of.

In the second half of the nineteenth century, they applied this technique to more and more stars and began to classify them according to their spectral lines. One of the first extensive catalogs of star spectra was published in 1890 and was based on the work of Williamina Fleming, one of the many women employed at the time to perform routine tasks at the Harvard College Observatory. These women, who could be paid much less than their male colleagues for this rather tedious work, had to assemble and classify data and make calculations. But Fleming, like many of her colleagues, didn't let this stop her from thinking about what she was doing. She developed a system of arranging the stars according to the relative amount of hydrogen observed in them: She labeled the stars containing the largest proportion of hydrogen with the letter A, the next group with the letter B and so on, down to N. Her colleague Antonia Maury had also developed a system, but one that was a lot more complex and intricate. In the end it was a third colleague, Annie Jump Cannon, who had been working at Harvard's observatory since 1896, who devised a compromise.

She adopted parts of Maury's and Fleming's systems, and then thoroughly reviewed everything again. She discovered that you didn't need all of Fleming's lettered classes to describe the stars, so she omitted some and rearranged the rest, retaining only the spectral classes O, B, A, F, G, K, and M. The fact that they're not in alphabetical order reflects the stars' temperature: O-type stars have a maximum surface temperature of more than 54,000 degrees Fahrenheit (30,000 degrees Celsius); next are the B-type stars, with a temperature of more than 18,000 degrees Fahrenheit (10,000 degrees Celsius)—both types emit a bright, blueish-white light. By contrast, A-type stars are noticeably less blue, and have a temperature of between 13,500 and 18,000 degrees Fahrenheit (7,500 and 10,000 degrees Celsius); they are followed by stars that emit a yellowy-orange light—types F, G and K—with temperatures of between 6,300 and 13,500 degrees Fahrenheit (3,500 and 7,500 degrees Celsius). At the far end of the scale are the cool, red M-type stars, whose surface temperature measures just 3,600 and 6,300 degrees Fahrenheit (2,000 to 3,500 degrees Celsius). To distinguish these types more easily from each other, Cannon added numbers to them: An F0-type star is thus a little hotter than an F1-type, which in turn is hotter than an F2-type, and so on, all the way to F9; and then we start again at 0 for the next type. In Cannon's catalog, therefore, the star HD 142 is a star that belongs to the hottest exemplars of the type G.

Over time, the spectral classes of stars have been further refined and extended, but we've kept to Annie Jump Cannon's original arrangement and designations, which is why you'll still find that odd, non-alphabetical classification of spectral types in modern astronomy. Generations of students have come up with mnemonics to help them remember the right order of the letters, but my favorite has to be "Our Bright Astronomers Frequently Generate Killer Mnemonics."

20

SIDERA MEDICEA
Not Stars, But Still Revolutionary

ON JANUARY 7, 1610, the universe shifted. At least our conception of how it's structured did. Ever since antiquity, people had believed that Earth was fixed to the center of the cosmos while the other celestial bodies moved around it. After all, we could see it with our own eyes: The heavens were revolving around Earth, and Earth plainly didn't move.

In the sixteenth century, however, Nicolaus Copernicus proposed a model of the universe that placed the Sun at the center, and turned Earth into merely one of several planets orbiting it. His hypothesis was rejected not only by the Church, but also by scholars: for, if it was true, how did it explain the Moon? In this new version of the cosmos, the Moon was still circling Earth, so that there were now two centers of movement: Earth revolving around the Sun, and the Moon revolving around Earth—it was wholly illogical.

But then Galileo Galilei appeared on the scene with his discovery of the Sidera Medicea. Dutch opticians had recently developed an instrument that could see better than the human eye; Galileo took it up and improved it, and then pointed the first ever telescope up at the skies. He saw things there that no one had seen before: mountains on the Moon; spots on the Sun; countless stars invisible to the naked eye. He could make out three such stars in the immediate vicinity of Jupiter, and over the following days it became clear that they were traveling across the sky in tandem with the large planet. More than that: They changed position—and then a fourth showed up. It looked exactly as if those celestial bodies were orbiting Jupiter.

Galileo wanted to name them Cosmica Sidera after his former student Cosimo II de' Medici, the grand duke of Tuscany—a gesture that he hoped would encourage financial investment in his future research. But Cosimo preferred the name Sidera Medicea, which would honor not only himself, but also his three brothers.

Of course, the Medicis' stars weren't actual stars. Galileo had found four of the moons of Jupiter. It was the first time anyone had seen a moon other than ours in the sky. More important, if Jupiter was also being orbited by moons, Earth couldn't possibly be the center of all movement—regardless of whether you looked at it in the context of the old or the new model of the universe. The discovery of those four moons of Jupiter was the first of many conclusive observations that eventually caused the downfall of our geocentric world view, and dethroned Earth from its seat at the center of the universe.

The name Sidera Medicea, however, didn't survive for long. The German astronomer Simon Marius, who'd discovered the moons of Jupiter independently only a day after Galileo, proposed to call them after characters in Greek mythology—but Galileo demurred and instead assigned them the Roman numerals I to IV. Yet when, in the nineteenth century, astronomers discovered more moons orbiting the other planets, things slowly became confusing, and they resorted once again to giving them proper names. Since then, the four moons have been known by the names that Simon Marius had originally proposed. Io, Europa, Ganymede and Callisto may not be stars, but they were certainly revolutionary.

21

HD 10180

Lots of Numbers, Lots of Planets

HD 10180 IS an extraordinary Sun-like star located about 127 light-years away in the Hydrus constellation. It's invisible to the naked eye, but the European Southern Observatory's telescopes watched it closely for six years. In 2010, the results were in: The star is orbited by at least six, perhaps seven and possibly as many as nine planets—more than have ever been found around any other star (except of course the Sun). Most of these planets are larger and heavier than Earth; in order of their distance from HD 10180, they're called HD 10180 c, HD 10180 d, HD 10180 e, HD 10180 f, HD 10180 g and HD 10180 h.

Is this really necessary? Why always these unwieldy, ugly combinations of numbers and letters? Where do these designations come from, and why do astronomers seem so reluctant to think up proper names for the stars and the other celestial bodies?

They don't because they can't—not because they're unemotional human beings lacking in creativity or a sense of romance, but because there are simply too many stars and planets in the universe, and we need a sensible way of organizing this multitude. In ancient times, it's true, stars were given "real" names—but not all of them, only the brightest. Sirius, for example, has a name that stems from the ancient Greek word *seirios*, which may have something to do with mythological tales of the sirens. In India, the star is called Makarajyoti, the Sumerians called it Sukudu ("heavenly arrow") and it's known by roughly fifty different names in different cultures.

This is the main problem with the names of stars: If the same object has more than one name, how can anyone know which one you're referring to? Until a century and a half ago, perhaps, it wasn't that much of a problem, but modern science is international, and efficient communication requires a systematic approach that can be readily grasped by everyone.

The first person to introduce a system for naming the stars was the German astronomer Johann Bayer in the early seventeenth century. He grouped the stars according to their constellations, then arranged them according to their brightness, and finally created designations for them from a combination of Greek letters and the Latin name of the constellation. Sirius, the brightest star in the Canis Major constellation, thus became Alpha Canis Majoris. In the eighteenth century, the British astronomer John Flamsteed introduced a similar system: He sorted the stars according to their positions in the constellations, and then simply numbered them, turning Sirius into 9 Canis Majoris.

Bayer's and Flamsteed's catalogs comprised only a few thousand stars. But in the course of time telescopes improved, and all those newly discovered stars in the sky had to be labeled, too. They were usually given just short designations consisting of letters denoting the relevant star catalog together with a combination of numbers and letters, which either pinpointed the stars' position in the sky or simply represented successive numbers in the given catalog.

The star HD 10180, for instance, home to several planets, is the 10,180th star in a catalog compiled by the Harvard observatory between 1918 and 1924 and named after the astronomer Henry Draper. It contains more than 300,000 stars, which is why we frequently find the abbreviation "HD" in star designations.

Of course, astronomers have continued to catalog ever more stars, and now there are numerous other catalogs, all with their own abbreviations. The most extensive collection available today was assembled by the Gaia space observatory: It lists 1,692,919,135 stars, which are known by the abbreviation Gaia DR2 followed by a nineteen-digit number.

Astronomers know, of course, that "real" names not only sound nicer, but are far more useful when talking to the general public about astronomy. That's why the International Astronomical Union established a Working Group on Star Names; it has tidied up and standardized the proliferation of names that have existed since antiquity, and some of the stars that used to have only "boring" catalog designations have now been assigned proper names, chosen from proposals submitted as part of a competition open to the public. The International Astronomical Union has so far officially recognized 336 star names. HD 10180, however, is still waiting in the queue.

22

TEIDE 1

A Star Gone Wrong

STARS RADIATE THEIR OWN LIGHT; planets do not. The reason for this is their mass: when enough material becomes tightly compacted, the enormous density and pressure cause it to grow extremely hot. The higher the temperature, the faster the atoms that make up the celestial body move, and when the temperature rises to more than 9 million degrees Fahrenheit (5 million degrees Celsius), hydrogen atoms move so quickly that they no longer bounce off each other when they collide, but instead bond into helium atoms. This process of nuclear fusion releases energy, and the object shines—and this is what we call a "star."

Any given star has to have at least 7 percent of the mass of the Sun, which is equivalent to seventy-five times the mass of Jupiter, in order to be able to emit light. Only when it has reached this mass can the hydrogen atoms inside it fuse. At least that's the case with normal hydrogen atoms—atoms whose nucleus consists of just a single proton. But many atoms also have variants, so-called isotopes; one of the hydrogen isotopes is called "deuterium" and has a nucleus made up of a proton as well as a neutron. Deuterium nuclei can fuse into helium atoms at much lower temperatures than "normal" hydrogen nuclei: An object with a mass thirteen times that of Jupiter will suffice.

Anything less than thirteen "Jupiter masses" makes nuclear fusion impossible, and a celestial body of that size is categorically not a star. However, if its mass is greater than seventy-five Jupiter masses—the point at which normal hydrogen atoms fuse into helium—we can be absolutely certain that it *is* a star. The interesting bit is the region in between, the gray area, where it isn't always entirely clear whether something is or isn't a star.

In 1962, the American astronomer Shiv Kumar took a closer look at this gray area. Not in the sky itself, but by using a theoretical model of these kinds of celestial bodies, which, he hoped, would enable him to predict what happens inside them. In his model, he encountered objects that shone a little like stars, but not very brightly and not for very long—and he realized that they did this because they were producing helium by fusing deuterium, rather than normal hydrogen atoms.

However, there is distinctly more normal hydrogen than deuterium in the universe. The reason that stars like the Sun can shine for billions of years is that they consist largely of hydrogen and contain only minute amounts of deuterium. An object that can't fuse hydrogen and has only a trace of deuterium available for fuel can shine, but only for a few hundred million years at most, and it releases a lot less energy in the process. Such celestial bodies merely shimmer a little to themselves for a while before their flame is extinguished and they cool down.

Shiv Kumar called these stars—which at the time existed only on paper—"degenerate objects" or "black dwarfs"; the American astronomer Jill Tarter later changed their name to the now commonly used "brown dwarfs." Nonetheless, for a long time scientists weren't sure whether an animal like this—something between a star and a planet—existed in reality. A brown dwarf is notoriously difficult to spot: It has to be smaller than true stars, hardly bigger than Jupiter; it shines weakly, if at all, and most of the light it emits falls within the infrared region of the spectrum, which is invisible to the human eye; and any slightly brighter and more massive brown dwarfs—though in theory easier to detect—are hard to distinguish from red dwarf stars, which only just about manage to achieve the required minimum of seventy-five Jupiter masses.

It wasn't until 1995 that we obtained incontrovertible evidence of a brown dwarf. Rafael Rebolo López from the Institute of Astrophysics of the Canary Islands and his colleagues were observing stars in the Pleiades star cluster 400 light-years away when they discovered Teide 1, an object roughly fifty times as heavy as Jupiter, whose surface temperature of about 3,600 degrees Fahrenheit (2,000 degrees Celsius) makes it distinctly cooler than the Sun.

Later, they showed that Teide 1 also contains the chemical element lithium—conclusive proof that we're dealing with a brown dwarf: If it were a true star, the nuclear fusion occurring inside it, as well as its high temperatures, would have destroyed the lithium long ago.

Since then we have discovered thousands of brown dwarfs all over our Milky Way, and we now know that there is no such thing as a clear-cut distinction between "stars" and "planets." The space between them is occupied by bravely glinting brown dwarfs. They may not have managed to grow into brightly shining stars, but that doesn't make them any less fascinating.

23

ALDEBARAN

Rendezvous in the Distant Future

IN ABOUT TWO MILLION YEARS, Aldebaran will receive a visitor. That's when the space probe *Pioneer 10*, which took off from Earth on March 3, 1972, will have covered the sixty-five light years that separate us from this star. However, it's unlikely that the probe will bump into any aliens in the area, because Aldebaran's life as a star is as good as over.

Aldebaran is in the Taurus constellation, and a worthy subject for observation in the night sky. You can easily see this brightly shining star with the naked eye, and also clearly make out its red color, which indicates its low temperature. Aldebaran is 3,600 degrees Fahrenheit (2,000 degrees Celsius) cooler than the Sun, but despite not being as hot, it shines much more brightly.

This is because Aldebaran is a "red giant." It has only 1.5 times the Sun's mass but is forty-four times bigger than our star. It wasn't always like that. Once upon a time, Aldebaran was a perfectly normal star and did what perfectly normal stars do: It fused hydrogen atoms into helium in its core—and in the process of this nuclear fusion, it released energy. But at some point the hydrogen in a star's nucleus runs out, leaving only helium for fuel, and in this short, final phase of a star's life, a star produces much more energy at its center than before and swells up.

This is one reason that the large, red Aldebaran is so highly visible. Another is its position in the firmament: It sits right next to the distinctive Pleiades star cluster, which it seems to follow, even as all the stars move together across the sky due to Earth's rotation. And that's how it got its name, which is Arabic for "the follower."

The Sun, too, will one day share Aldebaran's destiny. It, too, will swell up toward the end of its life and turn into a red giant. We don't yet know what this will mean for Earth, but what is certain is that the two innermost planets, Mercury and Venus, will be swallowed up and destroyed by the Sun. Earth may suffer a similar fate—or perhaps narrowly escape. For as the Sun balloons, it will inevitably lose a little of its mass; it'll grow so big that it can no longer hold on to the outermost layers of its atmosphere, which will disappear into outer space. The loss of mass will weaken the Sun's gravitational force, and its planets will shift a little farther away on their orbital paths.

But there's no need for us to worry too much about the future. All this will only happen in five or six billion years, and by that time Earth will have grown so hot that it'll be impossible for any life to exist anymore. And by then *Pioneer 10*, too—which set out in the direction of the stars to explore the remote regions of our solar system—will have long flown past Aldebaran.

However, we already lost all contact with the probe in 2003, which is a great shame. It would have been wonderful to find out what it might have seen on Aldebaran, because in 1998 we discovered a planet orbiting the red giant. Not a habitable one—it's a gas giant six times as heavy as Jupiter—but one that has evidently managed to survive the final phase of its star's life. And who knows: Maybe there's a moon, or a space station, orbiting the planet out there, occupied by some kind of being that could tell us what it looks like when a star dies . . .

24

WISE 0855-0714

All Alone in the Universe

WISE 0855-0714 ISN'T A STAR. But we don't quite know what to call objects like it. In the past, we used terms like "planetary-mass brown dwarf," "Y dwarf" and "planetary-mass object," as well as "nomad planet," "orphan planet" and "loner planet." The last of these is perhaps the most accurate: The object WISE 0855-0714 is an object three to ten times the mass of Jupiter, moving through space at a distance of 7.5 light-years from the Sun—but it's all alone, without a star to orbit.

That's why it's hard to describe it as a planet, because in astronomy the word normally denotes only celestial bodies that orbit stars. But nowhere in the vicinity of WISE 0855-0714 have we found a star to which it might belong. It's a loner—if not alone in its fate. It's estimated that there are about 400 billion free-floating nomadic "planets" in our Milky Way, almost twice as many as there are stars in our galaxy.

Their existence comes as no surprise. Ordinary planets, such as Earth and the other seven that make up our solar system, are created from a large disk of gas and dust that has formed around a young star. Over time, the material in this disk accumulates to form bigger and bigger objects, until at some point planets emerge. But we know from numerous computer simulations of these processes that a lot more planets are created than survive.

When our solar system was young, it probably contained a few dozen planets. But there wasn't enough room for all those celestial bodies—they exerted gravitational forces of attraction on each other, their orbital paths were anything but stable and there were numerous collisions. The Moon was created 4.5 billion years ago during one such collision between Earth and another planet. Earth survived the cataclysmic event, but the other, Mars-size planet wasn't so lucky. Some of the early planets fell into the Sun, others were hurled out of the solar system during near-collisions and what was left at the end of this chaotic beginning were the eight planets we know. The rest were destroyed—or are now wandering, alone and without a star, through space.

What happened here also happened everywhere else that planets came into being. All over the universe, young celestial bodies were exiled from their systems during the chaotic processes of their formation. It therefore comes as no surprise that there is such a vast number of vagabond planets. What is much more surprising is the fact that we're able to discover them at all.

Our discovery of WISE 0855-0714 was made possible by the fact that it's so close to us. Only the stars of Alpha Centauri, Barnard's Star and the two brown dwarfs of the Luhman 16 system are closer to the Sun. A nomadic planet may not have a star to illuminate it, but it has retained sufficient heat from the time of its formation, which it radiates into space in the form of a very weak infrared light. If a telescope can observe this radiated heat, it can also observe objects like this, and that's exactly what the space telescope WISE did in 2013.

Incidentally, we needn't be concerned that one of the many vagabond planets will collide with Earth on its route through the Milky Way. There may be lots of them, but there's also plenty of room in our galaxy. Within a volume of 100 cubic light-years there is, statistically speaking, a probability of just two roaming planets. The chance of a collision is therefore so minute that we can safely ignore it. They're lone wolves, yes—but entirely harmless.

25

WOLF 359

The Battle for Earth

JUST EIGHT LIGHT-YEARS from Earth there's a small star with the designation Wolf 359. A fleet of forty starships once decided the fate of our planet there. They prevented its invasion by extraterrestrial forces only by a whisker, and with a considerable loss of life.

All this happened in the year 2367, and not in the real universe, but in the world of *Star Trek*. In two episodes of *Star Trek: The Next Generation*, Wolf 359 was the site of a battle between Starfleet and the Borg. The complete assimilation and extinction of mankind could only just be thwarted, and since then the star called Wolf 359 has been famous among science fiction fans.

In the real world, this celestial body first appeared in 1919, in a catalog by the German astronomer Max Wolf. A few years later, the 359th star in his list turned out to be the Sun's immediate neighbor. Only the stars of the Alpha Centauri system and Barnard's Star are closer to us than Wolf 359, but you can't see this small and feebly shining celestial body in the Leo constellation with the naked eye. The reason that astronomers are interested in it is the variation in its brightness, which can give us an insight into its turbulent inner life. The rest of the world, or at least that part of it gripped by the adventures of the starship *Enterprise*, is probably more interested in the fictional space battle, during which none other than Captain Jean-Luc Picard himself was arrested by the Borg (if you don't know who he is, you're probably not a science fiction fan).

Astronomy and science fiction have long inspired each other: Wolf 359 alone appears in more than a dozen influential books, films and computer games, and a multitude of other real stars and planets have provided the setting for fictional stories. They include *Perry Rhodan*, the most successful and longest-running science fiction series of all time, which has charted the adventures of US Space Force Major Rhodan in 3,000 volumes (and counting) since 1961. In one of the first books, the hero and his comrades land on Venus; and in the fictional world of the series Venus is a warm, densely vegetated jungle planet, inhabited by all kinds of beasts, including huge, dinosaur-like animals.

When this story was written in 1961, we were already fairly sure that there aren't any dinosaurs on Venus. But we didn't yet know what things actually look like on our neighboring planet. Space exploration had only just begun, and the first probe landing on Venus was still years away. We knew that Venus is about the size of Earth, and a little closer to the Sun, but it's shrouded by a thick layer of cloud, which means that we can't see its surface from Earth. Although there was already some evidence in 1956 that the planet's temperatures are extremely high, in 1961 it was still thought plausible that it could be a planet like Earth, only a little warmer. Whatever we thought of *Perry Rhodan*'s jungle planet in those days, though, was literally science fiction—mere speculation about what might be possible, based on current scientific knowledge. That this speculation later proved to be entirely wrong (we now know that Venus's temperatures exceed an extremely inhospitable 750 degrees Fahrenheit/400 degrees Celsius) isn't particularly tragic. After all, that's precisely the role science plays in this game: Just as science fiction is inspired by science, the scientific imagination is sparked by the conjectures of science fiction.

Many scientific careers have started with *Perry Rhodan*, the USS *Enterprise* or other fictional adventures. Science, for its part, provides science fiction heroes with a stage for their experiences. This fruitful interaction between fantasy and reality will hopefully produce further discoveries about real and imagined worlds. After all, we all share the same mission—"to boldly go where no man has gone before."

SN 1990O

Dark Energy—an Unsolved Puzzle

THE WHOLE THING sounded relatively simple: 13.8 billion years ago, the universe started to expand. But given that the universe contains lots of matter, there has to be a limit to how far it can expand: The gravitational force exerted by all those trillions of galaxies with their billions of stars means that the expansion has to slow down over time—they tug at the expanding universe, which then loses momentum.

As I said, it sounded relatively simple—because we didn't yet know the extent to which gravitational force has already slowed down the expansion. It was this question that astronomers wanted to get to the bottom of in the 1990s. But how to go about it? How do you measure the rate at which the universe expands? Even more important, how do you measure the rate at which the universe expanded in the past?

For this, you first need enough type Ia supernovas, which is what we call the explosions that occur when a white dwarf briefly returns to the world of the living. A white dwarf is what remains of a small star, like the Sun, when the fuel that powers its nuclear fusion process has run out. It collapses under its own weight, creating an extremely dense celestial body, which then spends the rest of its life doing nothing much more than cooling down.

Unless it was once part of a tight double-star system, that is. In which case, material from its still-existing fellow star can transfer to the white dwarf, and thereby increase its mass. When the mass has exceeded a particular limit, nuclear fusion spontaneously and dramatically kicks in again. The white dwarf begins to shine again, much more brightly than before, and with such intensity that its life ends in a violent explosion.

The practical aspect of this special kind of supernova, from the astronomers' point of view, is that the maximum mass a white dwarf can reach before it explodes is always the same. The supernovas are thus always identical. How bright they are when seen from Earth depends only on how far away they are, and their great intensity means that we can observe them even at a vast distance—even when they occur in other galaxies.

There's something else you can do with the light from the explosion, besides calculate the distance. Just as the pitch of an ambulance siren changes as it passes us, the light emitted by a light source that's moving away from us also changes. The principle is the same, except that in this case it's a light wave, not a sound wave, and instead of the pitch it's the color of the light that changes. The more quickly the star is receding, the more marked the change.

We can therefore tell how far away these supernovas are, and how quickly they're moving away from us. However, a large part of this motion isn't a real movement of white dwarfs through space but is produced by the expansion of the universe. What's more, the greater the distance at which we observe an object, the longer it has taken for its light to travel through the cosmos to reach us, and the older the image we see. This is how remote supernovas can tell us the rate at which the universe expanded in the past—and by observing a number of such events at various distances, we can work out how much the expansion of the cosmos has slowed down over time.

That's why, in the 1990s, two independent research groups went on their separate searches for supernovas. The first was found in 1990, in an unassuming galaxy in the Hercules constellation. Its light had taken more than 400 million years to get to us. This first supernova, designated SN 1990O, was followed by a few dozen others in both nearby and remote galaxies.

The conclusions reached by the two research groups turned the model with which we'd been working entirely on its head, because their data showed that the expansion of the universe isn't slowing down. Far from it: Its expansion is accelerating. The finding was utterly unexpected, but it was reached independently by two different teams and has been confirmed several times since then. Something, then, is causing the expansion to speed up, but what? It's a question that continues to baffle astronomers to this day. They've fittingly called the phenomenon "dark energy" (which, incidentally, has nothing to do with "dark matter"). In the meantime, its existence has been confirmed beyond any doubt, and its discoverers received the Nobel Prize in Physics in 2011. The reason for the acceleration, however, remains unknown. It currently looks like there could be a kind of energy present in the "empty" space. As the universe has become bigger, it contains more space than before, as well as more energy, and it may be this energy that is propelling the expansion. But nobody knows why, or where it comes from. For now, the riddle of dark energy remains unsolved.

27

ALGOL

The Demon Star

ALGOL IS A STAR of ill repute. It's located ninety light-years away and is visible to the naked eye in the Perseus constellation, and has had many different names over the years. In ancient Greece it was called Gorgonea Prima, after the gorgons, those mythological monsters with snakes instead of hair growing out of their heads who turned anyone who looked at them into stone. Its Arabic name is *ra's al-ghul* ("demon's head"), later shortened to Algol. Algol used to be called the Devil's Star, or Ghoul Star, and in the Middle Ages was thought of as one of the unlucky stars. But what was it that ruined its reputation? Why, of all the stars in the sky, is Algol considered the monstrous one?

Because it breaks the rules. The other stars in the sky shine quietly and constantly to themselves; yet Algol appears to change, growing brighter and darker, so markedly in fact that we can clearly observe it even without technological aids. In the course of just three days, its luminosity fluctuates so much that it must have caused people sleepless nights 2,000 years ago.

After all, back in those days they had only a vague notion of what actually happens in the sky. They didn't know the true nature of the stars, and thought the heavens were part of a mysterious mythological world; many believed that celestial events constituted a message from the gods. The study of the stars gave them a glimpse of the future and allowed them to forecast significant events: When a catastrophe was to be expected, and when war, death or devastation would vanquish humankind. And a star like Algol—which behaves so very differently from all the other stars, which never rests but shines now more brightly, now more darkly in the sky—doubtless not only attracted people's attention but must have greatly unsettled them.

Algol is no longer a mystery to us. There are no demons or devils on it engaged in nefarious deeds, but there is something else: It's a triple-star system, two of whose stars are orbiting each other at close range. One of them is very large and a hundred times brighter than the Sun; the other noticeably dimmer. When, from our point of view, the two of them are positioned precisely next to each other, that's when the maximum amount of Algol's light reaches us. When one of the two stars obscures the other, the total amount of brightness we perceive decreases.

This is a "variable star" or "eclipsing binary"—a star whose luminosity varies not because of any processes taking place inside it, but simply because it consists in fact of a pair of stars that take turns to blot each other out. This phenomenon can be observed again and again elsewhere in the universe, and Algol has become godfather to an entire class of such celestial bodies: Algol variables. Incidentally, there is evidence to suggest that its odd behavior disconcerted mankind as long as 3,000 years ago. An Egyptian papyrus discovered in 1943 shows a calendar of lucky and unlucky days, a kind of "fortune calendar," and a closer examination of this calendar has shown that the lucky and unlucky days alternated every 2.9 days. This closely matches the period during which Algol's brightness changes. It may of course only be a coincidence, but it's quite possible that the star was seen as a kind of "metronome of doom" even then. And if you've been burdened by notoriety for that long, it's rather hard to get rid of it again.

28

POLARIS
One of Many

THE POLE STAR, currently also known as the North Star, isn't the brightest star in the sky, even if people keep saying it is. In the list of the brightest stars, it comes in a paltry forty-seventh. It owes its prominence merely to its position—and to the fact that Earth wobbles.

The Pole Star, also called Alpha Ursae Minoris or Polaris, is situated 430 light-years away in the constellation of the Little Bear, where it is indeed the brightest star. If you inspect it through a telescope, you can see that it's actually a triple-star system. The star that we can see so easily with the naked eye is a hot giant, which shines 2,000 times more brightly than the Sun; it's accompanied by a dwarf star, and the two of them are orbited by the third star of the ensemble.

What makes it so exceptional, though, is its position in the sky. If you extend the axis around which Earth rotates once every twenty-four hours from its northern end up into the sky, you'll hit almost exactly the spot where the Pole Star sits. The "celestial north pole" is the center around which all the stars seem to revolve in the course of a night. In truth, of course, they do no such thing, but remain where they are. But because Earth rotates around its axis, and we with it, it looks to us as if the stars circle the celestial pole—and thus the Pole Star, which appears to be the only immobile star in the sky.

Polaris is therefore very useful if you want to know where north is, but don't happen to have a compass at hand. Its practical significance is also reflected in its various old names: The

Anglo-Saxons, for example, called it *scip-steorra*, "ship's star," because navigators would orient themselves by it on the high seas—which is why it's also known as the "lodestar" (from *lád steorra*, "leading star"). Among Indian astronomers it used to be known as *dhruva*, meaning "fixed" or "immovable" star.

But although the Pole Star is for us the epitome of steadfastness and considered a dependable navigational aid, we can't in fact always rely on it, because Earth's rotational axis isn't stable. The Moon and the Sun's gravitational forces pull at it, and over time it traces a little circle in the sky. Right now, it happens to be directed almost exactly at the spot where the Pole Star is, but in the fourth century BCE the Greek geographer Pytheas of Massalia described the celestial north pole as free of stars, and rightly so: In those days, Earth's axis was pointing in a different direction. Ancient Chinese astronomers gave the star known as Kochab today the name 北極二 (*bei jí èr*), which roughly means "second star of the North Pole," and indeed, two or three thousand years ago this star was much closer than Polaris to the celestial north pole.

Even before that, people had identified the star Thuban (also known as Alpha Draconis) as the Pole Star. Between 3942 and 1793 BCE, this giant star in the Draco constellation was the one closest to the celestial pole, even closer than the Pole Star is today.

And if we go back even further, we'll find stars like Tau Herculis and Vega playing the role of pole star—and they'll do so again, one day. Because of the vacillation of Earth's axis, the celestial north pole will continue to coincide with the Pole Star well into the twenty-second century, but then its role will slowly be taken over by stars such as Gamma Cephei or Alderamin, until, in 12,000 years, it'll be Vega's turn again. Earth's axis draws its circles over a period of about 25,700 years, which is how long it'll be before the Pole Star once more plays the part for which it became famous among us. However, though we can rely on Earth's movement, we can't possibly predict whether there'll be anyone left here to gaze at the skies.

29

TYC 278-748-1

The Asteroids' Shadow

HOW BIG IS A STAR? Very big, that much is certain. But astronomers, of course, would like to know exactly how big. There are large stars and small stars, and if we want to understand them properly we should determine their size as accurately as possible. At first glance, we can only see them as spots of light in the sky, and a straightforward measurement is only possible with the help of immense telescopes—and then only for the stars closest to us. Yet astronomers are accustomed to approaching things creatively when it comes to getting hold of the data they need.

The star with the catalog number TYC 278-748-1 is a perfect example of this. It's in the Virgo constellation, and until May 22, 2018, it had been behaving fairly innocuously. But on this day, it was obscured by the 55-mile-wide (88 kilometer) asteroid Penelope. This was nothing new in itself—there's no shortage of stars and asteroids in the sky, and down here on Earth we can observe one of the small lumps of rock in our vicinity traveling past one of the stars in the solar system about once a week.

When this happens, however, a star doesn't extinguish all at once, like a lamp that's suddenly been switched off. Though it appears to us as no more than a dot, it is after all an object with a certain extent. It therefore takes a little time before the asteroid has covered it completely. How much time passes between first contact and complete darkness depends on the speed with which the asteroid moves, and of course also on the size of the star.

But there's an optical effect that prevents the straightforward measurement of a star's size during occultation. Starlight bends. It's a phenomenon that always occurs when a light source meets an obstacle: Its light is diverted and spreads out in directions that would otherwise be blocked off by the obstacle. In the case of an asteroid, it means that the asteroid doesn't cast a clearly defined shadow; instead, the center of the shadow is surrounded by a radiation pattern of light and dark waves. This initially makes it impossible for us to measure with any degree of accuracy how long it takes for the star's light to vanish. What we can do, though, is work out the diameter of the light source from the way in which the pattern changes as the asteroid crosses the star.

We can only do this if sufficient data is available to us. You have to measure the brightness of the disappearing star a few hundred times per second, and for that you need telescopes that can rapidly register changes in luminosity. In this case, astronomers repurposed VERITAS, the Very Energetic Radiation Imaging Telescope Array System at the Fred Lawrence Whipple Observatory in Arizona. These instruments are normally used to measure so-called Cherenkov radiation, which occurs, for example, when gamma rays hit Earth's atmosphere. Some stars emit this type of radiation just as they emit all kinds of other electromagnetic waves, so it isn't any wonder that astronomers are interested in it. But it isn't easy to observe it, because Earth's atmosphere doesn't allow gamma rays to pass through—which is a good thing, because they can do substantial damage to our body tissues and DNA. But the atmosphere itself also constitutes a disturbance for astronomers, who have to send gamma-ray detectors into space in order to observe the sky directly at gamma-ray wavelengths.

As is so often the case, however, astronomers have once again come up with an indirect method to solve the problem.

When gamma rays meet air molecules, they form particles such as electrons, which move very fast. So fast, in fact, that they can overtake the light in the atmosphere, which spreads a little more slowly there than it would in a vacuum. Only in a completely empty space does light travel at the maximum speed of 186,282.397 miles (299,792.458 kilometers) per second; as soon as it passes through air, water or another medium, it slows down. If other particles present in such a medium travel faster than the light, they produce the visual equivalent of a sonic boom: a bright flash. And this can be seen from Earth with the help of special telescopes—from which we can work out how much gamma radiation has hit our atmosphere, and where.

But the scientists observing the star TYC 278-748-1 through the VERITAS telescopes on May 22, 2018, weren't interested in stellar gamma radiation. Neither would the gamma-ray telescopes have been any good at taking "beautiful" pictures of the star. They can, however, detect even feeble light, and that's precisely what they were required to do that day. The star's brightness was measured more than three hundred times a second, and its diameter calculated with extreme precision from the fluctuations in its luminosity during the occultation: It's exactly 2,173 times as large as the Sun, and thus the smallest star ever measured. They now hope to make improvements to the telescope network and to their observational methods, so that in the future we'll be able to measure the sizes of many other stars just as precisely.

To establish the size of a star by means of asteroids and repurposed gamma-ray telescopes, astronomers had to think outside more than just the one box. And there's hardly another scientific discipline as accomplished in this as astronomy.

30

SS LEPORIS
To the Roche Limit

BRAM STOKER, STEPHENIE MEYER, *Van Helsing* and *Underworld*—for a long time now, vampires have helped to produce literary and cinematic blockbusters. But vampires exist, in a manner of speaking, in astronomy, too, although here they are considerably larger and more brutal than their cultural prototypes. While one undead bloodsucker consumes a few sips of blood, the cosmic vampire SS Leporis sucks up more than 132 quadrillion pounds (60 quadrillion kilograms) of material every second. This is quite a feat; to stomach such an abundant meal you have to be a fully grown star, and have a correspondingly voluminous victim at your disposal, consisting of something worth sucking.

You can find a pair like this in the sky, in the not at all bloodthirsty-sounding Lepus ("hare") constellation. In the binary system SS Leporis, the two stars orbiting each other have a somewhat one-sided relationship. One of them is large and cool, the other small and hot, and the small one has over the course of time appropriated almost half the larger one's mass.

It isn't usually that easy to take something away from a star. Stars are big, and with their enormous mass, and the gravitational force that goes hand in hand with it, they hold on to their material with everything they have. But for a star to be a star it has to shine. The energy it requires for this is produced in its interior, by means of nuclear fusion; and when the radiation subsequently travels from the center of the star to its surface, it presses against the star's matter. The longer a star lives, the hotter it becomes, and the greater the pressure from its radiation—which is why old stars swell and become ever bigger.

At one point, the star will have expanded so much that its gravitational force can no longer hold on to its outermost layers. The point at which it does this is called the "Roche limit" (after the French astronomer Édouard Albert Roche), and if a second star happens to be in the vicinity, that star can then appropriate the escaping layers of matter with its own gravitational pull.

For a long time, people thought that the same was the case for SS Leporis, too. But when, in 2010, they looked more closely through the telescopes of the European Southern Observatory, they realized that the situation there is a little more complicated. The large star may have swollen up considerably, but it hasn't done so enough to have reached the Roche limit. It should still be able to hold on to its mass.

But stars don't just calmly shine away; great eruptions occur on their surface, in the course of which material is flung out into space. We call this loss of material "stellar wind," and it appears to be particularly strong for the larger half of SS Leporis. And because the small "vampire star" is immediately next to it, it's in a position to suck up everything its fellow blows into space. It can only acquire the larger star's material because the latter is handing it over voluntarily.

The relationship between the stars of SS Leporis is therefore less like that between a classic bad vampire like Dracula and his unwilling victims, and more akin to those modern teenage fantasy stories, in which girls offer up their blood to worldly vampiric youths.

31

L1448-IRS2E

Star Under Construction

IT'S A LITTLE FUTILE to ask which star is the youngest of them all. New stars are formed all the time and everywhere in the universe. Even in our Milky Way, a few new ones are added each year. But to watch a star being born isn't very easy. Having said that, the discovery in 2010 of the object identified as L1448-IRS2E might just be such an astronomical stroke of luck. It was observed in the Perseus Molecular Cloud, a gigantic collection of gas and dust 850 light-years away—a promising start, because it is exactly regions like this one that produce new stars.

Left to its own devices, such a cloud of gas doesn't do much at first. It's in a so-called state of hydrostatic equilibrium, which means that while the cloud wants to collapse under its own weight, the atoms of gas inside it are in constant motion, colliding with each other and exerting a pressure that counteracts the gravitational force.

For something to change about this state of equilibrium, one of two things has to happen. Either the cloud's mass exceeds a critical value and then the pressure of the gas can no longer inhibit the gravitational force, and the cloud collapses. Or the cloud is disrupted by an external factor, for example another star passing close by and increasing the cloud matter's density with its own gravitational force.

Either way, the cloud eventually implodes. How far it gets depends on how quickly it manages to get rid of its energy. During the cloud's collapse, the energy of its motion is transformed into heat energy, making its center hotter and hotter.

This heat radiates outward, and so long as thermal radiation is able to penetrate the gas mass, everything is all right. The core becomes denser, but not hotter. At some point, though, it becomes so dense that the radiation can't pass through it anymore. Then the core heats up, and the increase in temperature accelerates the gas atoms. When a fresh equilibrium between the gravitational force and radiation pressure has been reached, the cloud's collapse is suspended.

The thing that thus forms in the course of a few thousand years is still extremely cool—only a few dozen degrees warmer than absolute zero. It's called a "first hydrostatic core," and is still far from being an actual star. For that to happen, the core has to collapse even further and grow much hotter. Theoretical models tell us that these first cores have to exist, but it's difficult to observe them. They hardly shine at all and radiate only a little heat. They exist in this form for only a few hundred or thousand years before proceeding on their way, becoming first a "protostar" and then giving birth to a star proper.

Xuepeng Chen from Yale University and his colleagues were therefore delighted when they encountered L1448-IRS2E. The object was emitting too little radiation to pass for a protostar, but they could see gas streaming from L1448-IRS2E's center at high speed, which is only possible when a sufficiently dense core with its own magnetic field has already formed in the center of the imploding cloud: When material from the outer edges of the cloud then falls onto the dense core, parts of it are redirected outward by the magnetic field.

What they were looking at had to be a "first hydrostatic core," an object as yet a long way from being an actual star, but already a little more than just a cosmic gas cloud. L1448-IRS2E has successfully completed its first steps toward becoming a star—but there's a long way to go yet.

32

NEMESIS

The Sun's Invisible Escort

NEMESIS KILLED THE DINOSAURS—and it's only a question of time before the Death Star strikes again and triggers a second mass extinction on Earth. This dramatic statement was made by the astronomers Marc Davis, Piet Hut and Richard Muller in 1984. According to them, the catastrophic asteroid impact that sounded the death knell for the dinosaurs 65 million years ago was not a unique event. In fact, we should expect such a catastrophe on average every 26 million years; the culprit is Nemesis, the Sun's as yet undiscovered companion star.

It sounds like a madcap theory. But these men of science had thoroughly good reasons for making that statement: The paleontologists David Raup and John Sepkoski had previously suggested that we can tell from the fossil record, which goes back many millions of years, that mass extinctions are periodic, reoccurring every 26 million years. Muller and the geologist Walter Alvarez discovered a similar periodicity in the age distribution of impact craters on Earth. Earth is clearly being peppered with asteroids and comets at regular intervals for some reason—and that reason, they argued, is Nemesis.

It isn't easy to identify a mechanism that can cause collisions at periodic intervals. Collisions between Earth and small bodies are meant to be accidental and follow no recognizable pattern. But what if, wondered Richard Muller, the Sun were actually part of a binary system? The idea isn't as extraordinary as it sounds, because the majority of stars in the universe don't exist alone but have one or more companions. Our solitary Sun is an exception—or maybe not. Such a faraway companion star would revolve around the Sun over a very long time span and might in the course of its travels occasionally end up in the vicinity of the so-called Oort cloud. This is the name we've given to the outermost regions of the solar system, and there, around ten to a hundred thousand times farther from the Sun than Earth, billions of small clumps of rock and ice are making their rounds.

This "rubble" left over from the formation of the solar system doesn't normally get in our way; the celestial bodies stay out there where they are, and don't come near us unless a star like the hypothetical solar consort passes close by and whirls them up with its gravitational force. Some of them might then be diverted into orbits taking them toward the center of the solar system, and within Earth's reach. This would result in more asteroid impacts than usual, coinciding with the times when Nemesis wanders too close to the Oort cloud on its path around the Sun.

In that case, though, shouldn't we be able to see Nemesis? After all, every star shines, and if it's a close companion of the Sun's, shouldn't we have identified it by now? Not exactly. It simply isn't all that easy to determine how far away a star is, especially the small, feebly glowing specimens—which, moreover, you have to find in the first place. It's possible that Nemesis has already appeared on any number of astronomical images—but only as one unassuming light spot among many. And it would plainly be too laborious to immediately calculate the exact distance of every star we happen to find somewhere, just for the sake of it. At any rate, this is how things stood in the 1980s. Since then, however, we've built space telescopes that have determined the positions and distances of numerous stars. If Nemesis is really out there, we should have found it by now.

Of course, we may merely have been unlucky during our sky surveys. It's also possible that Nemesis shines more weakly than we think. Or it may be that we haven't discovered Nemesis because it actually doesn't exist. Today, we consider this to be the likeliest explanation by far—for a closer examination of fossils and of the age distribution of the impact craters has cast doubt on this recurring 26-million-year period. Nemesis has been unmasked as a phantom. And our Sun remains a solitary star, forced to fly through space without a companion.

NAVI

An Astronaut's Prank

WHEN AMERICAN ASTRONAUTS began their journeys into space in the 1960s that would end with the historic Moon landing in 1969, they needed to understand not only their spacecraft, but also the skies. Like all explorers of the preceding millennia, they, too, had to know how to use the stars to orient themselves, and so they took charts with them that had labeled images of the brightest stars and constellations. Alongside the familiar names of Altair, Sirius, and Procyon, they contained three names that don't actually exist in astronomy.

The names Regor, Navi, and Dnoces can be found on the NASA astronauts' star charts, but they weren't made up by any Arab or Greek scholars thousands of years ago. They were the fruit of Virgil Ivan "Gus" Grissom's imagination. Grissom was the third man in space after Yuri Gagarin and Alan Shepard, and a pioneer of the US space program.

The long hours of navigation training, which the astronauts spent studying under the open sky and in planetariums, had put Grissom in a playful mood, and he came up with the idea of naming three of the celestial bodies on the NASA charts after himself and his colleagues.

He gave the star Gamma Velorum, in the Vela constellation, the name Regor. Read backward, it is the astronaut Roger Chaffee's first name. Iota Ursae Majoris, a bright star in the Great Bear, became Dnoces, after Edward White—in 1965, he had become the second man after the Russian Alexei Leonov to leave his spacecraft and spend time floating in space; hence "second," written backward, became the second star's new name. Gus Grissom immortalized himself in the Cassiopeia constellation: The central star of the readily identifiable "Celestial W," actually called Gamma Cassiopeiae, became Navi, Grissom's middle name written backward.

What had started out as a joke turned into a tragic memorial to the terrible events of January 27, 1967. On that fateful day, White, Chaffee and Grissom were scheduled to complete a routine exercise for the Apollo 1 mission. They were supposed to test the command module which, it was hoped, would one day take man to the Moon and back again. They weren't planning to start up the rocket that day; the sole purpose of the exercise was to try out the spacecraft's systems.

Three and a half hours after the test started, a fire swept through the module—the precise cause remains unclear. The automated life-support systems detected a drop in the amount of oxygen in the spacecraft—which wasn't surprising, given that the oxygen was being consumed by the fire. What was fatal, however, was that the system reacted to this by increasing the oxygen level. The added oxygen not only fueled the fire, but also raised the pressure in the module. It was impossible to balance out the pressure, which meant that the hatches could no longer be opened from the outside.

White, Chaffee and Grissom burned to death in their spacecraft. The first mission of the Apollo program had cost the lives of three people. In memory of the three deceased astronauts, NASA decided to adopt the star names Grissom had made up on a whim for their star charts, at least unofficially. And when Neil Armstrong and Buzz Aldrin landed on the surface of the Moon two years later, on July 20, 1969, they had with them a map of the stars on which were recorded the names of their late colleagues.

34

14 HERCULIS

Heavy Metal Stars

THE STAR 14 HERCULIS is full of metals—but this isn't to say that it's a large iron sphere. When astronomers talk of "metals," they mean something different than the rest of us do. At first glance, the chemistry of astronomy is very much more straightforward than what we learned at school: There's no complicated periodic table of the elements with its more than 100 entries. There's only hydrogen, helium—and metals.

From the point of view of astronomy this isn't at all irrational, even if it must seem like it to us here on Earth. The overwhelming majority of matter in the universe consists of hydrogen and helium, which are the two elements that formed right after the Big Bang. The rest of the periodic table was forged only later, through the nuclear fusion that takes place inside stars.

Compared to what occurred immediately following the Big Bang, this process is very inefficient. Even the Sun, which, at the end of the day, has had 4.5 billion years to create new elements, is made up of 71 percent hydrogen. As for the rest of its mass, just over 27 percent consists of helium, and all the other elements together constitute the remaining 2 percent or so.

Yet the Sun is actually one of the stars that contain a relatively large number of those other elements. When it was formed, the universe was already nearly 10 billion years old. The stars born before it had already had plenty of time to produce new elements and, when they died in huge supernovas, scatter them in space. From the outset, then, the cosmic gas cloud from which the Sun emerged already contained more than just hydrogen and helium, and in sufficient amounts to enable the formation of planets like Earth. We and our world depend on these new elements, but from a cosmic perspective they are present only in minute traces, which is why astronomers continue to group them under the (admittedly somewhat misleading) heading of "metals."

A star's metallicity, then, describes the proportion of its mass that is taken up by these metals, and is always measured with reference to the Sun. According to that yardstick, 14 Herculis's metallicity is markedly greater than the Sun's. This small star in the Hercules constellation looks rather ordinary at first: Located almost fifty-nine light-years away from us, it's a little smaller, lighter and cooler than the Sun, and glows orange. But it contains at least three times as many metals as our star.

On the one hand, this is a sign that it must be a fair bit younger than the Sun, and 14 Herculis is in fact only 3.6 billion years old; on the other hand, this also makes it a useful candidate for the search for new planets. If the cosmic gas cloud of which a star consists contains many elements in addition to hydrogen and helium, the matter contained in the cloud can assemble more easily and quickly into largish lumps—and thus turn into planets. In addition, those heavy elements act like a shield, protecting the lighter hydrogen and helium atoms from the powerful radiation that emanates from a young star. As a result, the gas atoms aren't as effortlessly destroyed and shifted out into space by the radiation and can instead take part in the process of planetary formation. It also makes it easier for the large gas planets, such as Jupiter, to see the light of day.

In 2002, this theory was confirmed by 14 Herculis, when we discovered a planet in its neighborhood whose mass is nearly five times greater than Jupiter's—which, as we know, is the largest planet in our solar system. If 14 Herculis wasn't a heavy metal star, it would no doubt be impossible for a giant planet like that to exist in its orbit.

35

ALPHA CAPRICORNI

The Wellspring of Shooting Stars

WHEN YOU SEE a shooting star, you have to make a wish. There may not be any astronomical proof that it will come true, but there's certainly no harm in trying. And why not wish for lots more shooting stars while you're at it? Because from 2220 on, it'll be possible—and this time we can state this with a fair degree of astronomical certainty—to see a myriad of them in the sky.

The star Alpha Capricorni gives us a little preview of what we might expect—even if it isn't a proper star and, strictly speaking, doesn't really have anything to do with it. This bright light in the Capricornus constellation consists of two stars that happen to be located so close to each other that they look like a single star to us down here on Earth. And Alpha Capricorni is the point in the sky in whose vicinity the Alpha Capricornids meteor stream appears to originate.

Despite their name, shooting stars aren't stars. They are miniature lumps of rock only a tenth of an inch or so (roughly a few millimeters) wide, and you can find them as space dust everywhere between the planets of our solar system. When Earth meets one of these grains of interplanetary dirt, we see a shooting star. The speck of dust hits Earth's atmosphere with a typical speed of between 20 and 45 miles (30 and 70 kilometers) per second. During its high-speed flight through the atmosphere, it rips electrons from the shells of the atoms the air consists of; and when these now shell-less atoms recapture one of the liberated electrons and reattach it, they emit energy in the form of light, which we then perceive as a shooting star.

The whole thing takes place about 60 miles (about a 100 kilometers) above us and lasts only a few seconds. By then—due to the considerable frictional heat created by its collision with our atmosphere—the speck of dust will have burnt out, and the spectacle is over. On a clear night you can almost always see a few shooting stars, but on some nights you can see veritable showers of them. The reason for this is the source of the dust.

It's produced by asteroids and comets, the smaller celestial bodies in the solar system. Comets, especially, are bona fide polluters. When they get close to the Sun, the ice they contain thaws, turns into gas and surges out into space, dragging with it all manner of dust from the comet's surface. Asteroids also sometimes throw up dust in this way, though usually much less of it; however, when they collide with other asteroids, for example, or get close enough to the Sun that they break apart, they, too, can produce a large amount of dirt.

So asteroids, and especially comets, drag a trail of dust with them through space, and when Earth traverses one such sullied region of the cosmos on its route around the Sun, we notice a clear increase in the number of shooting stars, which all seem to fall to Earth from one and the same place in the sky. These showers recur at the same time every year, because it takes Earth precisely one year to revisit these dusty regions in its orbit. For example, on late November nights we can observe the Leonids, so called because they appear to come from the Leo constellation—but we actually have the comet Tempel-Tuttle to thank for them. So far, we have identified more than thirty such "meteor streams," which provide us with showers of shooting stars, some of them more striking than others, at different times of the year.

Asteroid 2002 EX12 is responsible for the Alpha Capricornids. Earth encounters the dust it has left behind around July 30 every year. The meteor shower isn't particularly breathtaking yet, but a close examination of the asteroid has shown that it must have created a huge amount of dust when it broke apart rather dramatically about 5,000 years ago. Earth isn't currently moving through the bulk of it, but the extent and position of the dust cloud will shift over time, and things will look different from 2220 onward. Astronomers have predicted that the Alpha Capricornids will grow into an extremely powerful meteor stream and produce more shooting stars than any other. Whether all our dreams will come true that night remains to be seen.

36

ANWAR AL FARKADAIN
The End of the Night

THE STAR WITH THE EVOCATIVE NAME of Anwar al Farkadain belongs to the constellation of Ursa Minor (the Little Bear). Its Arabic name means "the brighter of the two calves," but it's not actually all that bright. You can only see it in a very dark sky, something that is sadly becoming rare here on Earth.

Most people can identify the Big Dipper in Ursa Major. Its stars are very bright, and therefore easy to see. It's not as easy with Ursa Minor: Its most distinctive star is the Pole Star at its tail, whose light is still comparatively bright, but the other six stars shine more feebly. Anwar al Farkadain is the darkest of them.

A glance at the Little Bear is thus a good opportunity to assess the quality of the night. The darker it is, the more of its seven most prominent stars you'll be able to see. Far away from brightly lit conurbations and other light interference, the complete Little Bear will be easily and clearly visible, but even just a small amount of artificial illumination—scattered light from a distant village, perhaps—is enough to brighten the sky to such a degree that Anwar al Farkadain is outshone and vanishes. As you approach the bright lights of big cities, most of the other stars in the constellation also vanish. Only the Pole Star and one or two others in the Little Bear can then still be seen with the naked eye, and they, too, are invisible when we look at the sky from the center of a sprawling, blazing city.

Nowadays, this view of a washed-out, gray and starless sky is almost normal. More and more people live in or just outside large cities, and everywhere light not only illuminates homes, streets and billboards, but also radiates directly up into the sky. Most of us can no longer find complete darkness; if you want to know what a starry night actually looks like, you'll have to travel to remote deserts or sail the high seas.

Yet light pollution is increasingly becoming a problem not only for astronomy. Many animals and plants are disturbed by the lack of darkness, and for us humans, too, the disruption of the rhythm of day and night can have negative consequences for our health. The unnecessary irradiation of the sky is certainly a waste of money: A sensible lighting concept that illuminates only what needs to be illuminated, and only when truly necessary, could save a substantial amount of energy and money.

More than anything else, though, the end of the night is a cultural loss. For thousands of years, the starry skies were humanity's constant companion. As soon as the Sun set, it was dark; you could see the Milky Way unfurling its ribbon across the sky, and thousands of glittering stars that shone so brightly that you could make out their colors. The sky has influenced our myths and religions; it has inspired philosophy, poetry, painting—and science, too. With the ubiquitous availability of electricity and thanks to artificial illumination, we have gradually lost our muse, and with it an important part of what makes us the questioning, searching, storytelling and reckoning creatures we are. For millennia, a glance upward would impel us contemplate the universe and our role in it. That the bright lights of our civilization have ended up outshining this space for thought may be one of human history's most ironic punchlines.

37

SIRIUS B

The Future of the Sun

IN FIVE OR SIX BILLION YEARS' TIME, when the Sun ends its active life as a star, it'll become what Sirius B is today: a white dwarf. We first observed this special end stage of a star's development in 1862, thanks to Sirius B. But it took a long time for us to fully understand white dwarfs.

A star is only a star for as long as it can produce energy. The temperature inside it has to be high enough to enable atoms to move sufficiently fast that they can fuse together when they collide. This nuclear reaction releases energy, which it radiates outward. The pressure exerted by this radiation is the only thing that counteracts the star's gravitational force, which would otherwise cause it to collapse under its own weight. Without nuclear fusion, a star would be unstable.

In the center of the Sun, hydrogen is fusing into helium. This process has been going on for 4.5 billion years, and the Sun still has enough hydrogen left to carry on with it for another 5 to 6 billion years. At some point, though, the helium will become a problem. All it does is lie around in the core, unable to fuse: Helium is heavier than hydrogen, meaning that it requires higher temperatures to fuse. The more helium is produced, the less room there is for hydrogen at the Sun's core—which is the only place hot enough for nuclear fusion. Consequently, less radiation reaches the outside, the pressure inside the Sun decreases and gravitation gains the upper hand.

The result of this process is complex. The Sun collapses slightly under its weight, and the matter inside it becomes a little more compacted, which in turn heats up the core. The hotter it grows, the more of the outlying hydrogen can now be drawn into the nuclear fusion process. In the end, it even becomes hot enough for the helium inside the core to start fusing.

These fresh energy sources produce a much greater heat, which leads to a sharp spike in radiation, which is now powerful enough that, instead of continuing to implode, the Sun starts to expand. As it swells up, its outer layers stretch and thin out, but at the same time the Sun's core continues to collapse. Since it originally contained much less helium than hydrogen, the helium-burning phase ends after a few hundred million years, and the core is extinguished. Away from the center, the hydrogen can continue to fuse for a while—but eventually this ends, too. The Sun's outer layers will by then have long disappeared into space and dissipated among the stars; the core, meanwhile, gradually collapses from the inside out until all that's left is an object about the size of Earth, in which nuclear fusion no longer occurs.

This is the white dwarf into which the Sun will one day turn. A white dwarf is still immensely hot, and its material is extremely tightly packed: A single teaspoon of its matter weighs as much as several cars. A white dwarf may be tiny in comparison to other stars, but it nevertheless retains a large amount of mass, and thus continues to exert gravitational force on its surroundings.

This is how we were able to track down Sirius B. "Sirius" is our name for the brightest star in the sky, but in 1844 we noticed it acting somewhat oddly: It was wobbling, as if it was being orbited by another star, which was influencing Sirius with its gravitational force. Yet there was no star anywhere in the area that could have been responsible for this.

Only in 1862 did Alvan Graham Clark, grandson of the telescope maker Alvan Clark, discover, as he was testing a newly built instrument, that as a matter of fact there *was* a second star. They called it Sirius B (its brighter, easily discernible companion officially bears the name Sirius A)—and many more discoveries followed on the heels of this one. We've already found more than a hundred white dwarfs within a radius of sixty-five light-years, and many more even farther away. So when the Sun's life ends in the distant future, at least it won't be lonely.

38

IOTA CARINAE

The Cosmic Eye Needs Glasses

FOUR BRIGHT DOTS on a slightly fuzzy dark-gray background: This unspectacular image, unveiled to a somewhat unimpressed public on May 20, 1990, kicked off a new era in astronomy. It was the "first light," the very first image from the Hubble space telescope—which already had a long history.

The American astronomer Lyman Spitzer first came up with the idea of a telescope in space back in 1946. It was common sense, really: The turbulence in Earth's atmosphere inevitably interferes with astronomical observation, and a telescope placed in space, beyond our atmosphere, would give us an unimpeded view of the stars and show us what we otherwise wouldn't be able to see. In the 1940s, though, space travel proper hadn't quite yet started, and it was absurd to think of sending an entire telescope into space. The project consequently only came to fruition in the 1980s, and on April 25, 1990, NASA finally launched the telescope named after the famous astronomer Edwin Hubble into space.

For its first systems test, they had chosen a region in the Carina constellation. There had been much discussion at NASA about whether or not to invite the press to the premiere—raw astronomical images, especially those taken during testing, are pretty unspectacular. To produce color images, errors have to be removed and different shots combined, and a lot of work is involved in the production of a presentable picture. While the very first Hubble image was of great interest to science, in the eyes of the public there couldn't be anything more boring.

But then it very quickly got very exciting. What Hubble had captured was definitely sharper and better than anything taken from Earth. The brightest star in the picture was Iota Carinae; we can see this star clearly with the naked eye, and its brightness made it easy to study in Hubble's image, too. However, closer examination revealed that the image wasn't as sharp as it ought to be. Ideally, a star should show up as a spot of light; it may always look like a small disk in reality, but NASA had expected that Hubble would be capable of focusing at least 70 percent of Iota Carinae's light inside a very small area. Instead, its light was smudged across a region of sky about ten times bigger than that. This expensive and long-anticipated space telescope turned out to have blurry vision.

The reason for this turned out to be the very definition of irony: The measuring device responsible for identifying flaws in the telescope's mirror was itself flawed. They had to wait three years to repair Hubble. Once again, a space shuttle transported astronauts into space, who installed a corrective system—you might say they gave Hubble glasses. From then on, Hubble was able to see things as sharply as it was supposed to, and it at last became the revolutionary instrument scientists had hoped it would be.

It had been a fairly bumpy start for the Hubble space telescope. But over the following decades its observations justified the effort, and the images it has sent down to us have transformed our conception of the universe; during its decades of activity, more than ten thousand scientific papers have been written using the data it has collected for us. Moreover, Hubble has fundamentally changed what the cosmos looks like to us earthlings, and its stunning images of stars, galaxies and cosmic nebulas have now become part of our collective view of the universe.

That unassuming, gray picture showing the blurry star Iota Carinae was the start of the most successful career any telescope has ever had. Hubble has continued its operations far beyond its intended life span of fifteen years, and the telescope celebrated its twenty-fifth birthday in 2015. But the end is nigh. NASA has retired its space shuttle program, so the telescope can no longer be serviced, and its orbit no longer adjusted. In 2024, at the latest, it will draw close enough to Earth that the friction in our atmosphere will cause it to plummet toward Earth and disintegrate. Perhaps for that one, brief moment, it will shine as brightly as Iota Carinae, the star with which its story began.

39

SUN

The Lengthy Search for the Astronomical Unit

THERE'S ONE STAR that is more important to us than any other, even if we barely think of it as a star. The Sun is so crucial to our life and our culture that it's easy forget its true nature—but of course it, too, is a star, and the Sun only appears to be entirely different from the many other stars in the sky because it's so very close to us.

But how close is it, exactly? We have asked ourselves this fundamental question virtually since the beginning of history. Ancient Greek scholars—including Archimedes, Aristarchus of Samos and Hipparchus—tried to answer it more than 2,000 years ago, but without optical instruments and a proper understanding of the processes that actually occur in the sky, their calculations were so imprecise that they didn't even begin to approximate the true number.

In theory, it should be easy. All you would need to do is measure the parallax—the apparent shift in the Sun's position when observed from different places on Earth. The closer to us it is, the greater the parallax, and from that you can then work out its distance. In practice, however, the Sun's tremendous brightness makes it impossible to determine its exact position against the backdrop of stars.

However, you can also approach the matter indirectly. The relative distances of the celestial bodies in the solar system have been known since at least Johannes Kepler and Isaac Newton's time—after all, Kepler's third law of planetary motion states that the time taken by a planet to orbit the Sun depends on its average distance from the Sun. Measuring the length of a planet's orbit is straightforward, because it can be done simply by observing its motion. But to use it in conjunction with the law of universal gravitation to calculate the planet's precise distance requires you to know the planet's, as well as the Sun's, mass. But these were as yet unknown.

All they could work out were the relative distances—that is, the distance between one planet and the Sun expressed as a proportion of the distance between the Sun and another planet. Thus, for example, they knew that Jupiter is on average five times farther from the Sun than Earth—but not what this measured in miles. To work out the absolute value from the relative distances, they first had to determine the absolute distance between any two celestial bodies.

They tried to achieve this by measuring the parallax of the Moon and Mars—and the values they obtained in the seventeenth century were already slightly more accurate than those of the ancient Greeks. But then they chanced upon a rare opportunity: In 1761 and 1769, they would be able to watch Venus pass in front of the Sun. Exactly how it does this and how long it takes depends on from where on Earth the event is observed, and of course on the distance between Venus and Earth. If you watch the "transit" of Venus from several different places and measure its duration as accurately as possible, you can then work out the distance between Venus and Earth—and consequently all the other absolute distances in the solar system.

In 1761, then, astronomers the world over were eagerly anticipating the transit of Venus. They prepared to watch it in Siberia, northern Scandinavia, India, Europe and America, and even sent observers as far away as Tahiti. Once they'd collected the data and performed the necessary calculations, they finally knew the distance between the Sun and Earth: 93 million miles (153 million kilometers).

Later they discovered more accurate methods, and in the twentieth century we measured the distance between Earth and Venus directly and with a high degree of accuracy using radar reflection. Of course, we'll never be able to determine the exact value, because the distance between Earth and the Sun constantly changes; our planet's orbit sometimes takes us a little closer to it, sometimes a little farther away. That's why the International Astronomical Union ended its search for an ever-closer approximation in 2012. It has since labeled the distance as an "astronomical unit," simply defined as a unit of length measuring precisely 92,955,807.3 miles (149,597,870,700 meters).

40

NOMAD1 0856-0015072

Pluto's Belated Revenge

ON NOVEMBER 6, 2010, the star NOMAD1 0856-0015072's light went out. Only for a few seconds, though. It was an event that, under any other circumstances, no one would probably have noticed. The star is invisible to the naked eye, and even if you look at it through a telescope it's just one unremarkable spot of light among many other unremarkable spots of light. Before November 6, 2010, nobody had been much interested in it, but on that day astronomers optimistically pointed their telescopes at it in the hope that it would tell them more about a particularly contentious and fascinating province of the solar system.

On January 5, 2005, the American astronomers Mike Brown, Chad Trujillo and David Rabinowitz had discovered an asteroid in the distant outer regions of our solar system. Today, the asteroid is known as Eris. Its path around the Sun lies far beyond Neptune's orbit and can achieve a distance from the Sun that is almost a hundred times that of Earth. A few asteroids had already been discovered in that region, but this one was special: Initial data suggested that it was larger than Pluto. Larger, that is, than the celestial body that was at the time still listed as the ninth planet of the solar system.

Had Brown, Trujillo and Rabinowitz discovered a new planet? Many, including Mike Brown himself, insisted that they hadn't. Eris had been encountered in a region of the solar system that contains a multitude of asteroids, and where the existence of asteroids is wholly to be expected. Eris was moving and otherwise behaving like an asteroid should. The only unusual thing about it was its great size, which made it the subject of intense debate: If Eris was an asteroid and larger than Pluto, what, then, was Pluto? Also an asteroid? Or should Eris actually be classified as a planet?

Most astronomers had in any case long believed that Pluto would be much better categorized as an asteroid. Like Eris, it's located in a region notorious for these celestial bodies and moves exactly like them. It is also much smaller than the other planets in the solar system. The fact that it was classified as a planet upon its discovery in 1930 can probably be ascribed to a shortage of available data. According to the prevailing opinion among astronomers, however, now that we knew better we should correct our mistake.

In the end, they did just that. In the summer of 2006, following heated discussions, Pluto was stripped of its planetary status. Today, just like Eris, it's "merely" a large asteroid orbiting the Sun together with many others. However, in recognition of Pluto and Eris's impressive sizes—impressive, that is, when compared to other asteroids—astronomers created the new category of "dwarf planets," which now encompasses all the bigger asteroids of the solar system.

Most astronomers have accepted this new state of affairs. Yet a minority continues to demand that Pluto should have its planetary status restored, and when the telescopes beheld NOMAD1 on November 6, 2010, the dissidents had a new reason to consider their demands justified.

For NOMAD1's short-lived extinction that day had nothing to do with the star itself. The responsible party was in fact Eris, which could be seen passing directly in front of NOMAD1. During the transit, the distant celestial body's shadow moved as a slim line across Earth, and three observatories in Chile were lucky enough to observe this rare "star eclipse." From the occultation's exact duration they were able to calculate Eris's precise diameter: With a diameter of 1,445 miles (2,326 kilometers), Eris is slightly smaller than Pluto, whose diameter measures 1,475 miles (2,374 kilometers). The demoted Pluto was bigger, after all.

Nonetheless, this won't help it to regain its former status because it isn't only the size of a celestial body that matters, but its environment. Pluto may be large, but it's one large asteroid among many, and so it fails to be a prominent enough celestial body in its sector of the solar system. However, this needn't be fatal—astronomers study asteroids with just the same pleasure and passion as they study planets.

41

Z CHAMAELEONTIS
Too Soon for Black Dwarfs

"DOES THE CATACLYSMIC BINARY Z Cha Contain a Black Dwarf Secondary?" is the titular question of a paper published by the astronomers John Faulkner and Hans Ritter in 1982. The not entirely serious law of headlines named after the journalist Ian Betteridge ("Any headline that ends in a question mark can be answered by the word 'no'") doesn't bode well for this question. Which isn't surprising, given that the universe is still far too young to harbor black dwarfs.

In Faulkner and Ritter's defense, however, they hadn't set out to find a "black dwarf" as defined by today's astronomers. What they were actually looking for was "brown dwarfs"—i.e., celestial bodies that are larger and heavier than planets but don't have enough mass yet to constitute actual stars.

We first began to suspect the existence of such nearly stars in the 1960s. At the time, we called them "black stars" because they barely shine. But later the term "brown dwarf" asserted itself, and "black dwarf" was used to describe an entirely different object, namely those celestial bodies that we may actually find—but not until the distant future—in the double-star system Z Cha.

The binary star Z Chamaeleontis (to give Z Cha its full name) is 316 light years away, in the Chamaeleon constellation. It consists of two dwarf stars, a red one and a white one. Red dwarfs are the most diminutive of all the stars: They have much less mass than the Sun, the smallest of them not even approaching 10 percent of the Sun's mass. White dwarfs are smaller still, and don't actually count as real stars. They are the leftovers of a medium-size star that has breathed its last.

When a star has used up all its fuel, all nuclear fusion ceases. Before this happens, during the final phase of its life, the star casts off its outer layers until only the dead core remains. This core is now only the size of Earth. It's still very hot at first, but now all it does is cool down—unless it can get hold of more fuel from somewhere.

The white dwarf of Z Cha is in luck, because its red companion is very close to it: so close, in fact, that gas is continuously transferring from the red dwarf to the white dwarf. As this happens, nuclear fusion briefly and explosively sets in again, and the star suddenly flares up.

At some point, though, this, too, will end, at the latest when the red dwarf has perished, too—and white dwarfs unlucky enough to have no companion meet their end even sooner. The final remnant of what was once a star then continues to cool and darken, until it has lost all its heat and is as cold as the universe that surrounds it. That's when it turns into a black dwarf: a cold, dark and dead celestial body.

This process takes a while. A white dwarf may be small, but it's nonetheless a colossal mass of extremely hot matter, and we estimate that it takes at least a quadrillion years for a white dwarf to turn into a black one, and in some cases very much longer. With its 13.8 billion years, the universe is simply too young to have produced a black star yet.

But when, in countless eons, their time does finally come, we'll for once be able to answer Faulkner and Ritter's question *contra* Betteridge's law: Then, yes, the binary system Z Chamaeleontis will have a black dwarf.

42

HD 162826

The Sun's Long-Lost Sibling?

THE SUN HAS MANY SIBLINGS. They're in all likelihood distributed throughout the entire Milky Way. It looks like we may have found at least one of them: HD 162826—a star 110 light-years away from us, and too feeble to be visible to the naked eye. But in 2014, the Armenian astronomer Vardan Adibekyan identified it as a possible sibling of the Sun's from a list containing a good 17,000 stars.

When a star comes into existence, it isn't alone. The large gas clouds in which stars are formed have sufficient material for hundreds, even thousands, of them. Once such a cloud has collapsed under its own weight, what emerges isn't a single, isolated celestial body: The collapse occurs simultaneously in several places, and in the end a whole litter of new stars is born. These stars may all be formed in the same region, but during the chaotic phase of their birth they take on slightly different "peculiar velocities." Although they move through space together at first, they soon disperse everywhere in the Milky Way.

The Sun is 4.5 billion years old, and to find its siblings after all this time is quite a task. To discover a "solar sibling," as these stars are known in astronomy, you have to bear in mind two things: Firstly, a star like that obviously has to be the same age as the Sun. If it is, it can still only be our starting point, since we can't determine a star's age with absolute certainty—and in any case, it may just be a coincidence. That's why, secondly, the stellar sibling also has to display the same chemical makeup as the Sun. Like the cloud in which it was formed, a star consists primarily of hydrogen and helium, but alongside them are trace amounts of other chemical elements. Their proportion differs from cloud to cloud but has to be more or less identical in all the stars born of the same cloud.

On his search for a star with the required characteristics, Adibekyan eventually encountered HD 162826. It has almost exactly the same chemical composition as the Sun and is also 4.5 billion years old. What's more, it's a star of almost the same mass, temperature, luminosity and size as the Sun. This is far from inevitable: In the same way that human siblings can look very different from each other, so can solar siblings. A cloud can create large and small stars, brighter and dimmer stars. That HD 162826 might be not only a solar sibling, but a solar twin, is a real stroke of luck.

Stars like this are of immense interest to astronomers. For one thing, examining solar siblings will enable us to find out exactly how our solar system was formed back in the day. For example, the Sun's chemical makeup is distinctly different from that of most of the other stars in its vicinity. This means that it must have ended up in the Milky Way after its formation, because our galaxy differs significantly from wherever the Sun was born. If we knew more of its siblings, we might find out how this came about.

Not only that, but the region surrounding a solar sibling would also be a good place to look for inhabited planets. We know that life on Earth—or at any rate, the chemical building blocks necessary for life—came into existence very early on (i.e., at a time when the Sun was presumably still dwelling among its siblings). Asteroid impacts may have catapulted chunks of Earth into space, including its building blocks of life, and even hardy microorganisms. Perhaps some of them managed to reach the planets of other stars that were nearby at the time and have produced life there, too. Or it might have happened the other way around, and life has come to us all the way from the planets of another star. Whatever the case, from an astronomical point of view at least, a family reunion would be splendid.

43

40 CANCRI

A Rejuvenating Collision

THE STAR 40 CANCRI is situated in the Crab constellation, roughly 630 light-years away. It's a so-called blue straggler, and for a long time astronomers were bewildered by these curious objects. Because, by rights, they shouldn't exist.

In 1953, the American astronomer Allan Sandage was observing stars in the globular cluster M3. The objects in a cluster of this kind share a past: They all emerged at more or less the same time, from the same large cosmic gas cloud. However, although they share a birthday, they develop differently. The more massive stars are hotter and the nuclear fusion inside them proceeds at a faster rate, which is why their lives are shorter than those of smaller stars, which have a lower mass.

When you look closely at such a cluster, therefore, you'll see lots of stars that have already grown into red giants, meaning that they are in the final phase of their life and have already swollen up considerably. By contrast, the less massive stars are still living a regular life. The older the cluster, the more of its stars have had time to develop into red giants, and if you compare the number of giants with the number of those that still shine normally, you can work out the cluster's age with a fair degree of accuracy.

During his survey, however, Sandage came across an odd couple of stars. They were bright and bluish—an indication that they were very hot, and therefore also had considerable mass. Yet all the other stars in the cluster with a mass similar to these blue mavericks had long ago developed into red giants.

Given that the stars there all came into being at the same time, these two blue stars should have ceased to exist by now. Simply put, they were too old to be able to exist in the shape in which they nonetheless presented themselves to Sandage.

Today, we know the secret of their rejuvenation. It's a somewhat radical course of treatment: A star begins to swell up into a red giant at the moment that the hydrogen in its core—which it needs to fuel its nuclear fusion process—runs out. If it wants to carry on living, it needs fresh material to power the nuclear reaction. One way for it to get hold of this is by being part of a very tightly knit binary system. If its companion started swelling up sooner, and if the two of them are close enough to each other, the dying star's material can transfer to the other one, which grows and grows, and becomes brighter and hotter than before. And thus its new life as a blue straggler begins.

In the case of 40 Cancri, however, things must have happened differently, because there is no indication that it was ever part of a double-star system. A rather more brutal rejuvenation must have taken place; it appears to have collided with another star, and when they collided these two low-mass stars merged into a single high-mass star, which was then able to start afresh.

44

171 PUPPIS A
The Birthplace of Gold and Silver

YOU CAN SEE 171 Puppis A in the constellation with the endearing name Puppis ("poop deck")—but not without some effort, because it's an extremely weak star. We can't simply tell from its whitish light what the star consists of, without the aid of a telescope's more sophisticated artificial eye. In 2010, it was one of the seventy-one stars that Camilla Juul Hansen and her colleagues at the European Southern Observatory subjected to detailed scrutiny. They were hoping to find out where gold and silver come from.

Just after the universe was formed 13.8 billion years ago, the range of chemical elements it contained was still limited. There was a lot of hydrogen, somewhat less helium, very small amounts of lithium and even tinier amounts of beryllium. The rest of the periodic table's elements still had to be created by the nuclear fusion that occurs inside the stars, where helium atoms can merge to create carbon and oxygen, which in turn can produce heavier atoms, and so on, until we get to iron.

Iron is problematic for stars. Every element of the periodic table down to iron always releases energy during fusion. But if you want to fuse two iron atoms, you need an energy input. Therefore, the chain of nuclear reactions that creates new elements inside stars ends with iron. The heavier atoms, such as gold and silver, must therefore have emerged by other means.

To produce gold and silver, you need neutrons. They are the building blocks of atomic nuclei and have no electric charge. If you bombard an atom with neutrons, they can easily penetrate its positively charged nucleus—whereas protons, the other group of nuclear building blocks, have a positive charge and would thus be repelled by the nucleus. Once enough neutrons have deposited themselves inside the atomic nucleus, it becomes unstable and disintegrates. This sets off nuclear reactions, during which some of the previously deposited neutrons are transformed into protons to create new, stable atoms like gold and silver.

We can replicate this process under controlled conditions using a particle accelerator, but where in space can you find accelerators? We don't know for sure yet, but what astronomers can say with a degree of certainty is that there are at least two kinds. Inside the cores of old, large stars, the fusion of hydrogen and helium has already ended, and only the elements located in their outer layers continue to fuse. This can produce a large number of neutrons, which can then kick off the process required to create the heavier elements. Yet neutrons are also produced in large numbers when a star ends its life in a supernova. Because the formation of heavier elements proceeds slowly in the first case and rapidly in the second, we label them (a tad unimaginatively) "s-process" and "r-process," respectively. How big a part this plays in the production of elements is something we don't yet fully understand; nor do we know exactly where and under what conditions the processes can best take place.

What's more, Camilla Juul Hansen and her colleagues established during their examination of 171 Puppis A and the other stars that these are in all probability not the only feasible processes. Hansen and Co. inspected stars with a variety of masses and at different stages in their development to determine which heavy elements were present in each of them, and in what amounts. Of course, the data will vary depending on the stars' characteristics, but if heavy elements are always produced in the same way by means of the same processes, they should be present in the same ratio, though their quantity may differ. For example, if a heavier star produces more silver than a lighter one, then it should also produce more gold than a lighter one. In many specimens, Hansen and her colleagues discovered exactly such a correspondence. But in many others, they didn't—in those cases, the reverse seemed to be true: The more of one element was being produced, the less they found of the other.

In particular, the formation of silver can't be accounted for purely by r- and s-processes, and it's likely that there is a third process involved, about which we know only very little so far. Gold and silver thus seem to take different paths during their formation, and heavy elements are presumably being produced in other places and by other phenomena in the universe besides supernovas and old stars. We have already discovered a few of these, for instance the high-energy processes that occur when neutron stars merge to form black holes. We were able to study this phenomenon only with the help of the latest technology— by observing gravitational waves—so who knows what literally valuable discoveries are still waiting out there for us.

45

ALPHA ANTLIAE
The Sky's Toolbox

THE STAR ALPHA ANTLIAE is rather boring, with little to distinguish it from the countless other stars in the sky. All in all, it's an unremarkable entity, and its name might well be the most interesting thing about it. It's the brightest star in the Antlia constellation (hence "Alpha Antliae")—which is Greek for "pump." In English it's therefore also known as "Air Pump." But who had the ludicrous idea of giving the constellation such a prosaic name?

It was the French astronomer Nicolas-Louis de Lacaille, who in 1750 set off on an expedition to Cape Town to measure the distance between Earth and the other planets. When you look at a planet from different places on Earth—i.e., look up into space from different directions—the planet appears to shift slightly against its starry backdrop. From this shift we can calculate its distance, and the farther apart the terrestrial observation posts are from each other, the more accurately we can do this.

So while his colleagues scribbled down their observations in Europe, Lacaille went to Earth's southern hemisphere, where he built himself an observatory and watched not only the planets, but also around 10,000 stars. From this data he created an atlas of the southern skies, in which he also outlined the known constellations. But then he realized that there was still plenty of room for new entries.

Just like art, literature and philosophy, European science has been greatly influenced by Greco-Roman antiquity. The same was and remains true for astronomy. However, in ancient times it was obviously impossible for someone living in the Mediterranean region to observe the stars of the deep south, which is why they didn't show up in classical star catalogs. The Greeks and Romans couldn't tell each other stories about them or associate them with mythical figures like the winged Pegasus, or vainglorious queens like Cassiopeia. And the stories that the inhabitants of the southern hemisphere told about those stars scarcely interested the European explorers and researchers of the seventeenth and eighteenth centuries.

In any case, Lacaille decided that vast regions of the southern skies were sorely lacking in constellations, so he pulled some out of a hat. To appropriately honor the Age of Enlightenment, he named most of them after scientific instruments, and when we officially reclassified the skies in the twentieth century, we adopted these names. As well as an air pump (which in those days was employed more for research purposes than to inflate mere bicycle tires—which in any case hadn't been invented yet), there is now also a telescope, a draftsman's compass, as well as a ship's compass; there are constellations named after the pendulum clock, the chisel and the painter's easel, and in the southern sky you can in addition find an octant, a sextant, a set square, a microscope and even a chemical furnace.

All those instruments feel rather like uninvited guests intruding on the romantic assembly of heroes, gods and other mythological figures we are so accustomed to seeing in the sky. What are a pendulum clock and a set square, after all, when compared to Perseus battling heroically with the monstrous Cetus to save Princess Andromeda? All those are constellations we can see in the skies of the north—but what dramatic stories can you tell about an air pump or a chemical furnace? But in a way, Lacaille was right: As long as they're lying unused in a dusty corner, all those scientific instruments may seem quite dull, but the things we have found out about the universe with their help are more thrilling than any myth.

46

W75N(B)-VLA2

A Baby Star Loses Mass

BECOMING A STAR AIN'T EASY. However, to describe how a star is formed is easy: Take sufficient stuff and squeeze it together sufficiently hard; at some point, the whole thing becomes so hot that the atoms inside it start to fuse, and this nuclear reaction releases energy, which makes the star shine.

It doesn't really matter what "stuff" you use for this. Using this recipe, you can even make a bona fide cinnamon star, so long as you have enough cinnamon to pack into the pile. Except on Earth, though, cosmic cinnamon deposits appear to be few and far between, which is why real stars aren't made of the sweet stuff, but of large cosmic clouds that contain mainly hydrogen. When such a cloud collapses under its own weight, its insides grow hotter. This heat radiates outward, pressing against the gas in the cloud, and when the temperature has risen high enough to enable the thus increased pressure from the gas to balance out the gravitational force, the implosion stops.

What has formed in this way is by no means a star yet. It isn't even a so-called protostar like W75N(B)-VLA2, which really deserves a better name. W75N(B)-VLA2 is 4,200 light-years from Earth, and currently has a mass eight times that of the Sun. It's still putting on mass, but also losing some. It is this very loss of mass that so intrigued the Mexican astronomer Carlos Carrasco-González and his colleagues in 2015.

For W75N(B)-VLA2 to become what it is today, it had to continue the described process for a bit longer. In its "proto-proto" state, this star in the making might be better described as a large, warm cloud. As fresh gas continually descends into the cloud from its outer layers, the shock waves it produces heat up the core, which consists of hydrogen molecules, which in turn consist of two bonded hydrogen atoms.

This is the structure in which hydrogen is most commonly found, but if you put enough energy into this kind of molecule it breaks apart. All the heat building up inside the cloud's core is now being spent on splitting hydrogen molecules, and without a continuing supply of heat the gas particles can't move fast enough to keep up the pressure necessary to prevent the cloud from imploding further. So the collapse begins afresh, and the cloud continues its process of implosion. As it does so its temperature increases, until at some point there's once again enough heat to produce the necessary gas pressure to stall the implosion a second time. Only then does the cloud turn into a protostar.

For now, it remains shrouded in a large amount of gas and dust. The gas and dust continue to sprinkle down on the cloud, making it denser and hotter—however, all that material collects into a disk around the star before it can get sufficiently close to increase the star's mass. Yet a portion of the material is also flung back out into space; this is down to the magnetic fields that the protostar has developed, which propel some of the gas away from the star.

We don't yet entirely understand how exactly this happens. But W75N(B)-VLA2 allowed us to actually witness some of what we had till then merely suspected. In an old image taken of the protostar in 1996, we can see it expelling gas in the shape of a spherical cloud. In more recent images, taken eighteen years later, the cloud has clearly flattened out. This confirms our hypothesis that, in the course of the star's formation, the material in the cloud arranged itself into a kind of doughnut around the protostar. That's why the material expelled by W75N(B)-VLA2 only spread where the ring didn't obstruct it, rather than freely in every direction.

To become a proper star, W75N(B)-VLA2 still has to get rid of the remainder of all that surrounding stuff, which will keep falling onto it for the next tens of thousands of years, making it even denser and hotter—until its core temperature has become high enough to carry out actual nuclear fusion. Only then will the protostar have become a star; and only then—assuming that it's bright enough—will we see it shining in our sky.

47

HIP 13044

A Case for Astro-Archaeology

IT LOOKS THOROUGHLY ORDINARY. Granted, it's slightly larger than the Sun, but also only slightly hotter, and at a mere 2,300 light-years away as good as in our galactic neighborhood. But it's almost twice the age of our star, and HIP 13044 has seen a lot over the years. It arrived here from far away, having been born in another galaxy; as a youngster, it had to stand by and watch as its galaxy came too close to the Milky Way and was ripped to shreds. Ever since, HIP 13044 has gone round and round the center of the Milky Way, in the company of other homeless stars.

HIP 13044 is a case for astro-archaeology—the somewhat tongue-in-cheek designation for the search for galaxies that no longer exist. Just like everything else in the universe, giant galaxies move through the cosmos. But when two of them collide, it isn't the same as when two cars crash into each other. There's no loud bang, nothing explodes and strictly speaking there isn't even something crashing into something else. There's simply too much space between the stars in the galaxies. Rather, think of the two powerful star systems as penetrating each other, and in doing so affecting each other with their gravitational force. It takes billions of years for such a collision to run its course. The galaxies keep whirling about each other until, at last, they merge.

If one of the galaxies is significantly bigger than the other, however, the smaller one may also just be swallowed up. This is exactly what our Milky Way has repeatedly done: Smaller star systems that came too close were stretched out by its gravitational force as the Milky Way pulled the stars in those small galaxies farther and farther apart; all that remained was a stream of stars, which is now wrapping itself around our galaxy. The only indication that it was once an autonomous star system is the motion of its stars, which move in the same direction as each other, at roughly the same speed, and not on the same plane as the rest of the stars moving about the center of the Milky Way.

We call such groups of stars "stellar streams," and HIP 13044 belongs to a stellar stream discovered by the Argentinian astronomer Amina Helmi and her colleagues in 1999. This structure, known today as the Helmi stream, has wrapped itself around the Milky Way several times, and it's estimated that it contains up to a hundred million old stars.

One of them is HIP 13044, and it made headlines in 2010 when astronomers from the Max Planck Institute for Astronomy in Heidelberg discovered a planet orbiting this star. We already knew of a few hundred planets of other stars back then, but this was the first known planet belonging to a star born outside our Milky Way—the first extragalactic planet we've ever met, and potentially a superb opportunity for astronomers to gain revolutionary insights.

This caused huge excitement among astronomers, but then it turned out to be too good to be true: In 2014, it transpired that they'd made a mistake in the interpretation of the observational data. The planet doesn't exist.

But there is a plentiful supply of other stellar streams. We have identified nearly two dozen of them so far, and sooner or later we'll discover a real planet somewhere inside these streams—an extragalactic "fossil" that will supply us with all the data that the nonexistent planet of HIP 13044 still owes us.

48

KIC 4150611

We Need More Syzygies!

"SYZYGY" IS MY FAVORITE WORD in astronomy. Where else can you find a sequence of letters this immoderately dedicated to the back end of the alphabet?

But a syzygy isn't only interesting for its spelling. It's the term used to describe a situation in which three celestial bodies are in exact alignment with each other. Most of the time, the word is used with regard to the Sun, Moon and Earth. When the Moon is positioned directly between us and the Sun, we witness a solar eclipse, and the same is true in reverse: When Earth is located exactly on a straight line between the Sun and Moon, we witness a lunar eclipse. Sometimes, the word is used for the alignment only of the Sun and Moon—after all, solar and lunar eclipses don't happen every month. The Moon may orbit Earth once every twenty-eight days, but when Earth is positioned between the Sun and Moon, but not exactly in a straight line between those two celestial bodies, what we witness isn't a lunar eclipse, but a full moon.

Yet we can also come by syzygies deeper in space—for example, if we look at the star KIC 4150611. Strictly speaking, it isn't a star, but five stars. Unlike the Sun, which has to travel through space all alone, most stars have one or more partners. The majority are double stars—i.e., two stars orbiting each other; there are also multiple star systems with more than two players (the maximum we've encountered so far is seven stars), but they become rarer the more stars they contain.

We know of only a few dozen five-star systems like KIC 4150611, whose quintuplets, situated about 420 light-years from Earth, are particularly notable. If we're lucky and the stars of a binary system orbit each other on just the right plane, it's possible for those stars and Earth to be in syzygy. Or, to put it differently: for one star to regularly cross between the other one and Earth. Of course, because both stars shine this doesn't result in an eclipse, but depending on which of the two stars happens to be in front of the other, the total amount of light that reaches us will be greater or smaller. Such "variable stars" are significant because their changing light allows us to measure both their size and exactly how long it takes for them to orbit each other, and thus to calculate their precise mass. Coming by those values is usually a very complicated undertaking, and it isn't always possible to work them out with a great degree of accuracy; however, variable stars pretty much serve them up to us on a silver platter.

As seen from Earth, the five stars of KIC 4150611 can form multiple syzygies at the same time. The brightest star of the ensemble is regularly occulted by a very closely paired binary system. But the stars of this double act periodically also pass in front of each other. And the two remaining stars in the troupe also circle each other, with regular occultations, even as they and the three other stars orbit each other.

So there's a bit of a kerfuffle going on in KIC 4150611. What with it being so crowded, there presumably isn't enough room for planets there, but if perchance there were some kind of living creatures there, they would no doubt think very highly of the word "syzygy."

49

DELTA CEPHEI

Henrietta Swan Leavitt's Wonderful Stars

IF IT WEREN'T FOR HENRIETTA SWAN LEAVITT, we would have no idea how big the universe is. Nor would we know this without the star Delta Cephei, which was central to the American astronomer's work. It's 887 light-years away in the Cepheus constellation, and easy to discern with the naked eye. However, it wasn't until toward the end of the eighteenth century that we realized that its brightness changes periodically. Nearly a hundred years later, Henrietta Swan Leavitt commenced her scientific work at Harvard.

She was engaged there as a computer. Back then, the term didn't describe the machines most of us have at home these days, but the people who executed mathematical calculations; and Edward Charles Pickering, the director of the Harvard observatory, mainly employed women to do this because in those days, just like today, you could pay them less than men for the same work.

Leavitt and the rest of the "Harvard Computers" were underpaid, but their scientific work was no less valuable for it. They were meant to perform "only" dull routine tasks—data analysis, star classification, that sort of thing—but no one could prevent them from thinking about the data they were handling.

As she worked, Henrietta Swan Leavitt noticed that there were stars whose luminosity periodically changed. She didn't understand why exactly this was happening, but she began to catalog them anyway. As she did, she realized that the brighter stars in her catalog changed their brightness more slowly than the less bright ones. You wouldn't ordinarily be able to do much with this piece of information, because what we see from Earth is only ever a star's apparent luminosity. As long as we don't know how far away a star is, we simply cannot determine whether a given star is really enormously luminous, or only appears to be so bright because it's very close to us. Determining its distance, too, is a relatively complicated procedure requiring a certain amount of rigamarole.

But Leavitt realized that it didn't matter in this case, because all the stars in her catalog were located in the Small Magellanic Cloud, a small satellite galaxy of the Milky Way about 200,000 light-years away, meaning that we can start from the assumption that all the stars in this dwarf galaxy are equally far away from us. Therefore, the brighter stars were actually shining more brightly than the darker ones—and thus in 1912 Henrietta Swan Leavitt was able to establish the so-called period-luminosity relation.

She did this using a particular group of stars called the "cepheids"—after Delta Cephei—that Leavitt had assembled in her catalog, and whose brightness is closely associated with the way their light fluctuates. We now know why they do this: It's the effect of the so-called kappa mechanism, which determines the capacity of a stellar atmosphere to absorb radiation. When an atmosphere is constituted in such a way that its transparency increases as it heats up, a feedback mechanism can develop, whereby the atmosphere becomes alternately more and less permeable to the radiation that emanates from the star's core; when this happens, the star, too, becomes alternately hotter and colder, which causes changes in its brightness. The pulsation rate also changes, depending on how bright and hot the star intrinsically is.

But even before we knew this, Henrietta Swan Leavitt's discovery of the fundamental relationship enabled us to accurately measure how far away the cepheids are. The periodicity of the changes in luminosity is easy to observe, as is the apparent brightness; and using Leavitt's period-luminosity relation, we could now work out the star's actual brightness and, subsequently, by comparing its apparent and actual brightness, also how far away the star is.

Henrietta Swan Leavitt showed us how to use the cepheids to measure the distance between us and the stars—thanks to this, we now know a little bit more about the true extent of the universe.

English American astronomer and astrophysicist Dorrit Hoffleit at the International Astronomical Union in Baltimore in 1988: She compiled the Yale Catalog of Bright Stars, listing all 9,095 of the stars that can be seen with the naked eye.

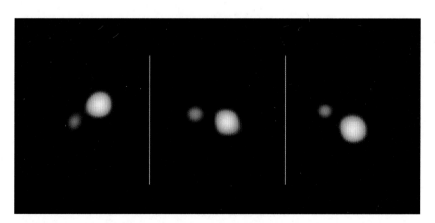

In the vampiric binary system SS Leporis, the two stars orbiting each other have a somewhat one-sided relationship. One of them is large and cool, the other small and hot, and the small one has over the course of time sucked away almost half the larger one's mass. These images capture the pair's clockwise orbit.

Sirius, a binary star and the brightest in the night sky, as seen by the Hubble Space Telescope: Sirius B, the tiny dot on the lower left, is a white dwarf—a glimpse of the fate of our own Sun, billions of years from now, when its nuclear fusion ceases and it shrinks to roughly the size of Earth.

This somewhat blurry, unspectacular image, unveiled on May 20, 1990, kicked off a new era in astronomy. It was the "first light," the very first image from the Hubble Space Telescope, aimed at part of the southern constellation Carina.

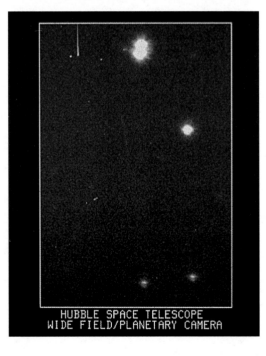

HUBBLE SPACE TELESCOPE
WIDE FIELD/PLANETARY CAMERA

American astronomers Annie Jump Cannon (left), who revolutionized the way we classify stars, and Henrietta Swan Leavitt, who measured the universe, standing together at Harvard College Observatory

British astronomer James Bradley, whose yearlong tracking of Gamma Draconis's movement became the first conclusive and indisputable proof of Galileo's theory that Earth orbits the Sun

American astronomer Robert Williams pointed Hubble as far as possible into the depths of space, at a tiny segment of supposedly empty sky, and photographed more than 3,000 galaxies. This galactic panorama became famous as the Hubble Deep Field, seen in part here.

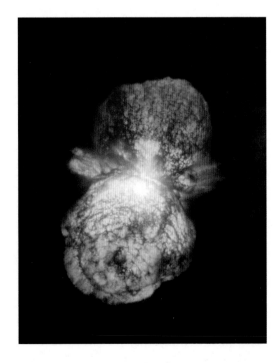

In the foreseeable future, very few stars in our vicinity might explode in a supernova; the massive Eta Carinae is one of them. Approaching its demise, it repeatedly initiates large explosions, which then hurl sections of its outer layers into space, as captured here by Hubble.

Astronomer and astrophysicist Cecilia Payne-Gaposchkin pioneered the use of stars' spectral lines to show that they consist chiefly of hydrogen and helium. Unable to graduate from Newnham College, Cambridge, because she was a woman, she moved to Harvard to write her dissertation, described by colleagues as "undoubtedly the most brilliant PhD thesis ever written in astronomy."

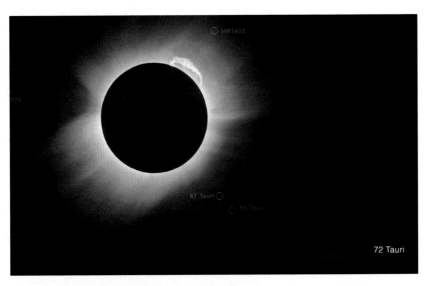

A copy of the photographic plate taken of the solar eclipse on May 29, 1919: British astronomer Arthur Eddington observed that the star 72 Tauri had changed position in exactly the way Einstein's theory of relativity would predict.

The arrow points to star V1 in the Andromeda galaxy, tracked over days by the Hubble Space Telescope, showing it at its dimmest and brightest. When Edwin Hubble first spotted V1 in 1923, he used its pulsation period to calculate its distance from Earth, proving it was located outside the Milky Way—the first evidence that the universe extended beyond our galaxy.

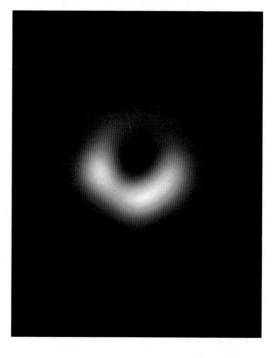

The first-ever image of a black hole, M87*, at the center of the galaxy M87: Really, this is its silhouette—black holes are by definition invisible—so what we see is a luminous disk of radio waves surrounding M87*'s so-called shadow.

The Nebra Sky Disk, the oldest known physical representation of the sky: On its iridescent green surface, the Sun, Moon, and stars are arrayed in gleaming gold. Of the original thirty-two stars on the disk, only thirty remain, including the group of seven known as the Seven Sisters, or the Pleiades.

2008ha

The relatively dim SN 2008ha supernova: That the life of a star could end in a bang that was more of a whimper was news to astronomers. Spotted by fourteen-year-old citizen scientist Caroline Moore, the youngest person ever to discover a supernova, her discovery thus provided astronomers with an entirely fresh insight into how stars die.

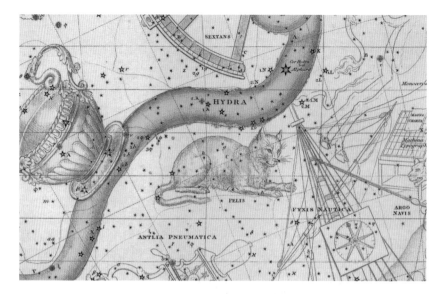

The nonexistent constellation Felis (Latin for "cat"), outlined by cat-loving French astronomer Jérôme Lalande in 1799 and adopted by various European star maps until 1888, shown here in Alexander Jamieson's 1822 *Celestial Atlas*

For thirty seconds on March 19, 2008, we were able to see with the naked eye farther into the universe than ever before, when the gamma-ray burst GRB 080319B's afterglow came into view. It was the most immense explosion ever recorded.

50

THE STAR OF BETHLEHEM
A Messiah's Status Symbol

THE STAR OF BETHLEHEM belongs to Christmas just like the execrable mulled wine they sell at Christmas markets. Unlike that alcoholic hot drink, though, this celestial body can at least justify its existence with a Bible reference. The Gospel of Matthew tells the story of the three "stargazers" who wanted to attend the birth of Jesus and made their way to the site of the big event using not GPS, but the stars: "And, lo, the star, which they saw in the east, went before them, till it came and stood over where the young child was. When they saw the star, they rejoiced with exceeding great joy."

From an astronomical point of view this is rather sparse information, especially because it's the only extant report of this star's materialization. It appears that not a single person witnessed this extraordinary apparition anywhere else in the world. Nevertheless, astonishing efforts have been made over the centuries to uncover a scientific basis for the phenomenon. The star is often represented as a comet with a sweeping tail rather than a star—an image we owe to the fourteenth-century Italian painter Giotto. He was presumably inspired by a real comet: Halley's Comet, which flies past Earth every seventy-six years, made a famous appearance in Giotto's time, too, and could be seen with the naked eye. However, at the time of the Nativity it wasn't anywhere near Earth, and as far as we know no other comet passed our planet at the time in question.

Johannes Kepler thought that the star of Bethlehem may have been a supernova—that is, a star that had ended its life with a colossal explosion when the fuel powering the nuclear fusion in its core ran out. When something like this happens close enough to Earth, it really is possible to see a "new star" in the sky for some time afterward. Such an explanation may in theory tally with the biblical account, but a supernova would have left a trace visible even to this day, and nowhere have we found the remnant of a stellar explosion fit for a supernova from the turn of the first millennium.

Another popular theory unceremoniously turns the star of Bethlehem into a conjunction of the planets Jupiter and Saturn. A "conjunction" is how astronomers describe the phenomenon of two celestial bodies seeming to be in exceedingly close proximity to each other. It was the Viennese astronomer Konradin Ferrari d'Occhieppo who first came up with the hypothesis, which he surprisingly explained in astrological, rather than astronomical, terms: Jupiter and Saturn are, he claimed, the astrological symbols for the Jewish people, and the sign of Pisces represents Palestine; so when Jupiter and Saturn meet in Pisces—which in fact happened around the time of the nativity— it should be read as a sign of the birth of the king of the Jews in Palestine.

Or not. Astrology isn't a science, and its criteria permit all sorts of things to be inferred from celestial bodies. Moreover, even when there is a particularly close conjunction, you can clearly see that it concerns two planets—not one star, as the Bible would have it. The stargazers, especially, would definitely have noticed the difference.

The real problem, though, with all these attempts to explain the star of Bethlehem, is the assumption that it actually existed. The Bible isn't a factual account, let alone a scientific publication. It merely pursued certain narrative structures, motifs and topics common at the time, and according to the most convincing theory—one based on historical and scientific insights—the star of Bethlehem is in truth a mere fiction: a cosmic "status symbol" with which Matthew wanted to underscore the significance of the nativity.

At the time, such motifs and narrative techniques were not uncommon in Roman historiography either. According to the Roman historian Suetonius, even Julius Caesar—who died just a few decades before Jesus was born—had "his own star": a brightly shining celestial body that appeared in the firmament after his death and was generally taken to be a sign of his apotheosis and newly attained divinity. And Matthew might well have thought that if Caesar had his own star, Jesus simply had to have one, too.

51

ARCTURUS
The Speed of Rainbows

ARCTURUS, WHICH IS IN THE BOÖTES ("ox-driver") constellation, is a giant star twenty-five times larger and two hundred times brighter than the Sun. Despite that, its mass is very small, barely greater than that of our star. This is down to the fact that Arcturus is already approaching the end of its life. The originally smallish star has swollen up to a mammoth size and has become a red giant. You can detect its red glow easily even without a telescope, and in springtime it forms part of the celestial Spring Triangle, together with the two bright stars of Spica in the Virgo constellation and Regulus in the Leo constellation.

What also distinguishes Arcturus is its speed. It's faster than almost any other bright star, traveling at a whopping 76 miles (122 kilometers) per second relative to our solar system. But there's no need to fear that it'll collide with us anytime soon, because its "radial velocity" is just a touch over 3 miles (5 kilometers) per second.

In astronomy, it's often useful to distinguish between the different components of a celestial body's speed, and to examine only that component of it with which an object is coming straight toward us: its radial velocity. So although Arcturus is racing through space at 76 miles (122 kilometers) a second, from our point of view its movement is a "sideways" one more than anything else, meaning that the star is flying past us slightly diagonally. Its distance from Earth is consequently decreasing by only 3 miles (5 kilometers) a second, and it'll move away from us again in just under 4,000 years.

A star's radial velocity is of great importance to astronomy, among other things because it can be determined quite simply by examining its light. We do this by separating starlight into its constituent parts—i.e., the different colors—using special optical instruments. This artificial rainbow—the "spectrum"—normally shows "spectral lines," which form when the gas atoms a star consists of block out part of its light. Every chemical element produces its own characteristic pattern of lines, and we can predict exactly where on the rainbow these will occur.

However, they only turn up where we expect them to if the star isn't moving in relation to us. If it is moving—regardless of whether away or toward us—the same thing happens as with an ambulance siren whose pitch rises or falls depending on which direction it's moving in; though here it's a case of light waves rather than sound waves, so instead of the pitch it's the frequency of the light emitted by the star (i.e., its color) that changes. Light waves radiated by an approaching source shift toward the blue end of the rainbow; when the light source moves away everything shifts toward the red end. As a result, the spectral lines end up in different places, too—and it is this shift that allows us to calculate the radial velocity.

The effect becomes particularly interesting when the star isn't just moving, but "wobbling." This can, among other things, be caused by the presence of a planet: For instance, as Earth revolves around the Sun over the course of twelve months, our planet's gravitational pull makes the Sun move a little, too; it describes a small circle, and from a distance it looks as though it spends half the year moving a little toward the observer before retreating again.

In a case like this, therefore, the shift in the spectral lines is periodic, now tending toward red, now toward blue; and from its periodicity we can work out the weight of the planet that's causing it, as well as how long it takes to orbit its star. This is the method astronomers used to discover the first planet of an alien star—and hundreds more such extrasolar planets since. However, to discover a planet like Earth, you really have to take extremely precise measurements. The change in the Sun's periodic radial velocity caused by our planet comprises 3.5 inches (9 centimeters) per second, distinctly slower than a leisurely stroll through the garden. Well, stars sometimes like to take it easy, too.

52

GAMMA DRACONIS
And Yet It Moves!

"AND YET IT MOVES!" Galileo is said to have mumbled obstinately in 1632 after having been forced by the Inquisition to recant the notion that Earth moves around the Sun. But this is only a legend, and just as false as the geocentric world view favored by the Church in those days, which placed unmoving Earth at the center of the universe.

During the seventeenth century, the idea that Earth orbits the Sun increasingly gained traction, and in the eighteenth you'd have been hard-pressed to find anyone who still harbored any serious doubts about it. However, there was as yet no direct proof of Earth's movement around the Sun. This was provided only in 1725 by the star Gamma Draconis—and in quite an unexpected way.

Gamma Draconis is a bright star belonging to the Draco ("dragon") constellation and is easily visible close to the celestial north pole. The British astronomer James Bradley subjected it to observation because he wanted to measure its parallax— the apparent shift in its position in the sky, which is due to the fact that Earth orbits the Sun in the course of a year, and we thus look at the stars from different directions at different times. From its parallax, you can calculate the distance of a star from Earth, which is something nobody had yet managed to do. Bradley wanted to be the first, and during his yearlong nocturnal observations he was indeed able to establish that Gamma Draconis moves.

But this movement wasn't the one that Bradley had actually wanted to see. The star was traveling in the wrong direction: In the course of a year, Gamma Draconis described a small ellipse in the sky, and Bradley had no idea why. At first he thought it was merely because Earth was wobbling a little, perhaps as a result of the gravitational force exerted on it by the Moon. In that case, though, a star visible on the opposite side of the sky from Gamma Draconis would have to precisely mirror its movement. Which, as an experiment showed, was unfortunately not the case.

It took a while before Bradley was able to explain the phenomenon. But eventually two things occurred to him: Firstly, light isn't infinitely fast. It takes time for it to get from one place to another, even if those two places are only the top and bottom end of a telescope. Secondly, Earth moves around the Sun. It does this even as the ray of light emitted by the star is itself in the process of traveling from the top end of the telescope' to the bottom end. If, at the time, Earth happens to be proceeding directly toward the source of the starlight, nothing unexpected happens. But if Earth is moving in another direction, then the telescope moves, too, and the ray of light arrives at the bottom end of the telescope with a small shift. This effect is more pronounced the greater the angle between the star's ray of light and the direction of Earth's movement. But this angle constantly changes as Earth moves around the Sun, which is why it seems to an observer like Bradley as if the star's light were traveling through the telescope from different directions at different times in the course of the year.

This effect—which Bradley was now finally able to explain—is called "aberration." And it was the first time that someone had proved conclusively and indisputably by observing the skies that Earth orbits the Sun. It would be nice to imagine that, having realized this, Bradley exclaimed "And yet it moves!" But, just like Galileo, he didn't.

MERAK

Bears in the Big Dipper

MERAK IS ONE of the seven bright stars in the Big Dipper. Nearly all of us know where to find this cluster of stars in the sky, and many people think that what they're seeing is a constellation. However, the Big Dipper isn't a constellation but a so-called asterism, that is, a distinctive lineup of stars to which we have assigned a particular significance, in addition to the official constellations.

Ever since we first consciously looked up at the sky and perceived the stars, we have also been grouping the stars into images and interwoven them with their myths. Which stories we would tell about them depended on when and where in the world we were, and the stars at the center of the narratives weren't always the same, either. "Official" and globally acknowledged constellations have only existed since 1928, when the International Astronomical Union divided the sky into eighty-eight distinct areas that we now describe as constellations.

That's how Merak and the other six stars in the Big Dipper ended up in the constellation Ursa Major; yet they aren't in themselves a constellation. They are comparable to the distinctive Pleiades cluster, an asterism that today forms part of the Taurus constellation. Some asterisms transcend the boundaries of their constellations: Deneb in Cygnus, Vega in Lyra and Altair in Aquila form the well-known Summer Triangle, which—unsurprisingly, given its name—we can see especially well when it's summertime in Earth's northern hemisphere.

Imaginative people can see other things, too. A teapot, for instance, in the Sagittarius constellation; a clothes hanger in the Vulpecula constellation (albeit only with a telescope or pair of binoculars); Kemble's Cascade, a line of twenty stars in the Camelopardalis constellation; or a fly in the Aries constellation. Give your imagination free rein, and there's no limit to how many you can find.

In a manner of speaking, Merak itself is part of a second asterism. Together with Dubhe, it forms the front end of the Big Dipper, and the two stars are sometimes also known as "the Pointers": If you extend the line between them in the sky about fivefold, you'll end up directly at the Pole Star, which indicates north.

But Merak has a few things to offer from an astronomical point of view, too. It's a large, hot giant star with twice the Sun's mass and three times its size, almost 9,000 degrees hotter (5,000 degrees Celsius) and more than seventy times brighter. Together with nearly all the other bright stars in the Big Dipper and Ursa Major, it belongs to the so-called Ursa Major Moving Group—a troop of stars that, as seen from Earth, all move in the same direction. This is because they actually have the same origin. About 300 million years ago, they were formed out of a huge gas cloud. As it moved through the Milky Way, this star cluster slowly disbanded, and although the stars are still moving roughly in the same direction as before, some of them are doing so a little more quickly than others. And if the stars of the Ursa Major Moving Group don't look to us as if they were part of a group, instead seeming to be situated far apart from each other in the sky, that's because of their (comparatively speaking fairly small) distance from us: The cluster's center is currently about eighty light-years away, and Merak, at seventy-nine light-years, is even a tad closer.

It will be quite some time before this pack of bears has disbanded to such an extent that we'll no longer be able to make out the Big Dipper, and before Merak and Dubhe can no longer serve as pointers to the north. If there are still any human beings looking into the sky in that distant future, however, they'll doubtless make out new images there, and be inspired to tell new stories.

54

GS0416200054

Creative Roads to Discovery

FOR 100 HOURS, the Hubble space telescope surveyed a very particular place in the sky: a tiny spot in the Ursa Major constellation where, as far as we knew at the time, there was nothing whatsoever to see. Yet there was method in the madness, because nowhere in the universe is there "nothing whatsoever." And if you're imaginative enough you may discover new things precisely in those places where, at first glance, there's nothing to see.

It was the American astronomer Robert Williams who first thought of closely examining this nothingness. At the time—in 1995—he was the director of the Space Telescope Science Institute, the organization responsible for overseeing the Hubble space telescope operations. Which was, from a scientist's standpoint, lucky for him, because you can only get a certain amount of observation time with Hubble, and not all who want to use it to explore the skies can do so. You have to submit an application, and only the best projects succeed in convincing the committee to allow you a glance through the telescope.

To gaze upon nothingness for 100 hours is an undertaking that would normally stand no chance of being approved, given the strictly limited observational time available. But because Williams was the director, he had a certain quota of observational time at his disposal to do with as he pleased. And what Williams wanted to know was how galaxies form and develop. To find out, he had to look as far as possible into the depths of space; the more remote an object is, the longer it will have taken for its light to cover the interval between it and Earth, and the further we can see into the universe's past—the place where you can observe galaxies in their infancy.

But you can do this only if the light from those distant galaxies is able to make its journey to us unimpeded, and if it isn't outshone by the other stars in the neighborhood. In short, where the sky is completely empty. Or looks empty: because, according to Williams, there must be galaxies all over the universe, and as long as nothing blocks your view and nothing disrupts your observation, when you look closely even an apparently empty region of the sky will be teeming with galaxies.

He turned out to be right. By the end of his observational campaign, Williams had photographed more than 3,000 galaxies in that tiny segment of sky he'd been looking at. This galactic panorama became famous as the Hubble Deep Field, and to date more than 500 scientific papers have been written about it. The supposedly empty sky has proved to be a veritable fount of knowledge about the young universe.

A few years later, Konstanze Zwintz from the Vienna University Observatory and her colleagues came up with a creative way of obtaining new insights without needing to be granted observational time with the Hubble telescope. Zwintz was interested in a particular type of fluctuation in the brightness of starlight, which is most easily and accurately measured using a space telescope, especially if you are able to observe a star for a substantial length of time. But what if you haven't been granted permission to use the telescope? Then you use the stars that Hubble has to observe anyway.

To capture images of the Hubble Deep Field, the telescope had to be oriented as accurately as possible. To do this, it referred to so-called guide stars, an unchanging set of stars whose positions are known. One of them has the designation GS0416200054, and during the ten days that the telescope spent observing the Hubble Deep Field, the sensors directing it kept this one (as well as other stars) almost continuously within their sights. All the data relating to the guide stars recorded during the telescope's orientation was freely accessible, and Zwintz and her colleagues scrutinized it for indications of brightness variations.

Their imaginative scheme unfortunately came to nothing this time, but Zwintz's team was at least able to demonstrate that the method is theoretically suitable for research of this kind, and that we can select guide stars more efficiently in the future by choosing specimens that not only point the telescope in the right direction, but also can in themselves generate new astronomical discoveries. When attempting to extract as much information as possible about the universe from the stars, the creativity of researchers knows no bounds.

55

PSR B1919+21

Smashed and on Its Last Legs

MATTER LOVES TO BE around other matter, and it's sometimes almost impossible to keep it away. Gravitational force knows only one direction—it always attracts—and if you amass enough matter in one place it gradually collapses under its own weight. What happens then depends on the forces capable of counteracting the collapse.

In an ordinary star, radiation pressure is one such force. Its matter becomes so compacted and hot that nuclear fusion can take place and begin to release radiation. This radiation pushes against the matter inside the star and prevents it from collapsing further. But when the nuclear fusion process eventually ceases, gravitational force takes over again. It increasingly compresses the atoms that make up the stellar matter, and the force with which it does this depends on the star's mass. In the case of a comparatively small star such as the Sun, the collapse ends once the atoms are packed like sardines: Quantum mechanics tells us that two electrons cannot be in the same place at the same time, and atomic shells are formed of electrons. As the collapsing star's gravitational force attempts to compress the atoms ever more, the electrons start moving more and more rapidly to and fro; this so-called degeneracy pressure suffices to delay the collapse, and the resulting object is called a "white dwarf."

However, if the original star has a mass greater than the Sun, degeneracy pressure isn't enough. Its gravitational force will be powerful enough to overcome it—it will, to put it simply, squeeze the electrons in the atomic shell into the nucleus. Or, to be more precise: The positively charged protons in the atomic nucleus and the negatively charged electrons interact with each other, turn into uncharged neutrons (one of the building blocks of atomic nuclei) and release neutrinos—extremely light elementary particles that as good as never interact with other matter.

As it transforms entire atoms into neutrons, the star's core becomes increasingly dense. At the same time, there are now countless neutrinos pushing outward at great speed. These highly charged particles contain a large portion of the energy released during the nuclear reaction and are now rapidly carrying it toward the star's surface. The star collapses; its outer layers are driven into the dense core, producing a shock wave that spreads outward and in the process rips the star's outer layers with it. The result is a powerful explosion called a "supernova"—which leaves behind it nothing but the very dense core of the star, now consisting exclusively of neutrons.

These "neutron stars" are only about the size of a city like Denver but have more mass than the Sun. If you were to squish 900 Great Pyramids of Giza into a teaspoon, you would end up with pretty much the same density as the material inside a neutron star.

Neutron stars form whenever big stars die yet are so small that you can barely see them directly. Indirectly, yes: Because of the star's extremely high density, its originally weak magnetic field also grows denser and more powerful. In addition, the star's rotational speed quickly accelerates—think of ice dancers spinning faster as they tuck their arms in. A neutron star can spin more than a thousand times a second on its axis, and as it does so its powerful magnetic field snatches nearby particles out of the gas cloud left behind by the supernova and hurls them into space at a great speed. In the process, these particles emit radiation; not light, but long-wave radio signals, which orient themselves by the direction of the magnetic field.

Like a lighthouse, then, a neutron star—in this case called a "pulsar"—sends radio waves into space with each rotation. We can receive the signals whenever the radiation "beacon" grazes Earth, and the pulsar then appears like a cosmic tracking signal emitting powerful radio pulses at regular intervals.

This phenomenon was first observed by the British astronomer Jocelyn Bell (now Dame Jocelyn Bell Burnell) in 1967. She was working on her radio telescope when she discovered a source of radiation that she labeled PSR B1919+21. Its unofficial nickname is LGM-1, short for "little green men," because astronomers thought this regularly transmitting radiation source so strange that, for a brief time, they actually considered the possibility that we were receiving signals from an alien civilization. But it didn't take long for them to discover the truth, which to their surprise turned out to be even stranger than the "alien hypothesis": because it looks like there are dead stars out there in space, sending radio signals through the universe.

56

CANOPUS

Blinding Brightness

CANOPUS SHINES MORE BRIGHTLY than almost any other star in the night sky. You can find it in the Carina ("ship's hull") constellation, but to do so you have to travel far into the south. It's impossible to see it in far-northern parts of the world, despite its brightness. It's one of only four stars (the others being Sirius, Arcturus and Alpha Centauri) whose "magnitude"—the term astronomers use to describe a star's brightness—is negative. That this value is less than zero for Canopus shows how old our science is, and how longstanding its traditions. And exposes its affinity for strange units.

The classification of the stars in the sky according to their apparent brightness can be traced back to ancient Babylonian scholars. They assigned the brightest stars to the first order of magnitude, those still just about visible to the sixth, and the rest to the four in between. The magnitudes allocated to the celestial bodies were later adopted by Greek scholars like Hipparchus, who used this system in the second century BCE for his star catalog. The stellar magnitudes were passed down through the centuries, but it was the British astronomer Norman Pogson who, in 1850, first attempted to formulate a precise definition.

In this, he was guided by the so-called Weber–Fechner law of sensory perception, which states that our subjective experience of the intensity of an impression is proportional to the logarithm of the intensity of the stimulus that triggers it. Or, put differently: If twice as much light reaches us from one star than from another, the star doesn't appear twice as bright to us—according to the system of stellar magnitude, every magnitude represents a roughly 2.5 increase in luminosity compared to the preceding magnitude. Pogson thus fixed the brightness of a first-magnitude star as being exactly a hundred times greater than that of a sixth-magnitude star. To create a reference point for his scale, he defined the Pole Star's brightness as 2.1 magnitudes.

However, we later discovered that the Pole Star's brightness changes a little over time and chose other stars to set the different orders of magnitude. That's why the brightest stars in the sky have now dropped out of the top end of the scale and been assigned negative magnitudes. Ever since, Canopus has shone with a brightness of –0.62 magnitudes. Only Sirius and a few of the planets we can see at night are brighter: Mars can reach a brightness of –2.91 magnitudes, Venus –4.6 magnitudes. Only the Moon and the Sun are brighter still: The latter achieves an impressive –26.73 magnitudes.

The fact that astronomy describes the brightest objects using negative numbers takes a little getting used to. The smaller the magnitude, the brighter the celestial body shines. In addition, the scale isn't linear, but logarithmic. Of course, we could have defined things differently—but it's how they did it in Babylon, it's how they did it in ancient Greece, and so it's how we still do it today. Astronomy is the oldest science in the world, and sometimes you just have to bow to tradition.

ETA CARINAE
A Leak in the Hull

THE STARS WE SEE in the night sky are actually no longer in the places where we see them, and many of them are no longer around at all. You'll often hear people make this surprising statement, which is as sensational as it is untrue. It may be true that a look out into the universe is also always a look into the past: Light, with its speed of 186,282.397 miles (299,792.458 kilometers) a second, is exceptionally fast, but the distances between the stars are far greater. Starlight takes decades, sometimes centuries, to reach us. We therefore perceive the stars only as they were at the time their light set off on its voyage to Earth, and when we gaze upon distant galaxies we might even be looking billions of years into the past.

But when it comes to the stars that are visible to the naked eye, the case isn't quite as pronounced. These stars reside in our own Milky Way galaxy and are at most a few thousand light years away—which means that their light has been traveling toward Earth for "only" a few thousand years—a drop in the ocean compared to a star's typical life expectancy of billions of years.

It is also true to say that the stars move around the Milky Way, but given the enormous distances involved we hardly notice it. We can register their movement with the help of highly accurate measuring instruments, but this doesn't change anything about their positions as we perceive them with the naked eye; unless, that is, you observe them very closely for a few decades or centuries.

Thus the stars we see are exactly where we see them, and we can be reasonably sure that all of them still exist. Naturally, they won't stay around forever. Every star has to die sometime and will stop shining when the fuel that powers its nuclear fusion process has run out. But observational data allows us to estimate with a reasonable degree of precision how many years a star has left to live. And there are only a very few stars in our vicinity that might expire in the foreseeable future.

Eta Carinae is one of them—a powerful specimen of a star located in the constellation with the pretty name Carina, meaning "ship's hull." It has one to two hundred times the mass of the Sun and is thus one of the most massive stars that can possibly exist. A star of that size is correspondingly hot, and shines very brightly: Eta Carinae is five million times more luminous than the Sun. Yet the hotter a star is, the shorter is its life span, and although Eta Carinae isn't 3 million years old yet, it has already burned up a substantial portion of its fuel.

Its behavior, too, indicates its approaching demise: It's repeatedly initiating large explosions, which then hurl sections of its outer layers into space. During one such eruption in the mid-nineteenth century, it briefly became the second-brightest star in the night sky; then it dimmed again, and by the start of the twentieth century it was no longer visible to the naked eye. Since then, it has again become brighter, and it's only a question of time before the next violent eruption happens. Soon it will explode completely and vanish from the sky.

When the "ship's hull" springs its stellar leak, however, we probably won't be able to watch it happen: I meant "soon" in its astronomical sense—i.e., in a few decades or centuries, perhaps a little longer.

58

ALPHECCA

A Colorless Jewel in the Celestial Crown

ALPHECCA IS THE BRIGHTEST STAR of the constellation with the handsome name Corona Borealis, the "northern crown." That's why it's often also called Gemma, Latin for "jewel." Nevertheless, Alphecca is no glittering precious stone. It's a white star seventy-five light-years from the Sun, and officially completely colorless.

Alphecca was one of the stars used by the American astronomers Harold Johnson and William Morgan in the 1950s to mark the zero point of the so-called UBV system, a photometric system for determining a star's brightness.

Until relatively recently, astronomers used only their eyes to survey the sky. This didn't change with the invention of the telescope; telescopes enabled us to see more and better than before, but in the end the light in the sky still shone on human eyes. Only when, in the nineteenth century, photography began to play an increasingly significant part in astronomy did things become a little complicated when it came to determining luminosity.

The reason is that a star's brightness depends partly on the color of the light we see. Our eyes always perceive a blend of all visible colors, and we don't distinguish between one brightness and another. However, a photographic plate is more sensitive to some colors of light than others, so the intensity of the pictured star can differ from its intensity as perceived by our eyes. That's why we always have to decide exactly which color of light we are interested in and allow only light of that specific color to be captured by the plate—and this is where the photometric system comes in: Having first set out the precise wavelengths of various colors of light, we then compare the brightness of different stars only within the relevant range.

Using the UBV system, we observe ultraviolet light at a wavelength of 364 nanometers (U), blue light at 442 nanometers (B) and yellow light at 540 nanometers (V), where "V" stands for "visual" because our eyes are best able to see in this range.

The intensity of a typical star's light differs within each of these wavelength ranges, which we can then use to create so-called color indices. If a star shines more brightly in the B filter than in the V filter, that means that it's emitting more blue than yellow light, and the color index resulting from these two luminosities (B-V) is negative. (Which sounds wrong, but isn't: In astronomy, the number used to describe a star's brightness grows smaller the brighter the star.) In this case, the color of the star is blueish. Conversely, a star that emits red light has a positive color index. This system of color indices allows astronomers to describe the color of a star in precise mathematical terms. But every system needs an absolute zero, because a white star is necessarily equally bright in every color of the spectrum and should have a color index of zero.

To calibrate their system accordingly, Johnson and Morgan searched for very bright white stars to use as the zero point of their photometric system. Among them was Alphecca—which thus became a colorless jewel in the celestial crown.

59

BARNARD'S STAR

A Controversial Fast Bowler

FIXED STARS. WE STILL HEAR this term used to describe the stars in the sky. Yet stars are anything but fixed. That's just how they look to us—and if we look really closely, we'll discover that even this isn't true. Barnard's Star is the perfect case in point.

We're dealing here with an unremarkable star, so unremarkable in fact that you can't even see it with the naked eye. However, Barnard's Star is very close to us—a mere 5.9 light-years away—and only the three stars of the Alpha Centauri system are closer to the Sun. But what makes Barnard's Star special isn't its proximity to us, nor is it its appearance—it's a red dwarf star, like the vast majority of stars. But Barnard's Star clearly demonstrates that "fixed stars" are anything but fixed.

At first glance, it really looks as if the stars don't move. The motion we observe is only an apparent one: Because Earth rotates on its axis, the whole sky, including the stars, revolves around us in the course of a day. But the relative positions of the stars remain unchanged, so when we look at the sky today we see the same constellations that the stargazers saw in the days of old.

It's therefore understandable that some people think the stars don't actually move. Understandable, but wrong. Everything in the universe moves, and the stars are no exception. The stars in our Milky Way are all moving around the center of the galaxy—it takes the Sun roughly 220 million years to do this, while other stars take less, or longer. And the Milky Way itself is also moving through the universe, as are the countless other galaxies and their stars.

Yet the stars aren't moving around the center of the Milky Way with the same regularity with which the planets orbit the Sun. The stars sort of wobble around, exert a gravitational force of attraction on each other and change each other's course— this is called "proper motion." As seen from Earth, therefore, the stars change position over time, but you need to look very closely to see them do it. The stars are so far away that their movements can't be tracked with the naked eye, and you would have to observe the skies for thousands of years to notice the difference.

However, toward the end of the eighteenth century, using decent telescopes and accurate instruments, we managed to measure the movement of the stars. Then, in 1916, the American astronomer Edward Emerson Barnard discovered that no known star moved as quickly across the sky as the one named after him today: Barnard's Star, also known as Barnard's Runaway Star, is currently racing toward us at 91 miles (146 kilometers) a second. Yet it won't ever run into us; in just under 10,000 years, it'll fly past Earth about 3.75 light-years away.

In the 1960s, Barnard's Star once again caused much excitement when the Dutch astronomer Peter van de Kamp declared that he'd found a planet almost twice the size of Jupiter in its vicinity. He claimed that the star jiggled about as it moved, which could only be explained by the gravitational pull of a nearby planet. It would have been a sensational discovery, because astronomers had been searching for extrasolar planets for a long, long time. But not everyone was convinced, and a closer examination of the data in fact showed that the star only ever seemed to wobble when van de Kamp had just serviced and adjusted the telescope.

Independent observations were unable to confirm van de Kamp's hypothetical planet, and in recent years, thanks to the availability of even more accurate instruments, we've ascertained that it would be quite impossible for a large planet to exist there.

In 2018, to everyone's surprise, researchers did actually encounter a planet in its vicinity—but one much smaller than the one stipulated by van de Kamp, with only a shade more than three times Earth's mass. Van de Kamp's hypothesis may have been flawed, but he would probably have saluted the discovery.

60

DENEB

Cecilia Payne Fathoms the Stars

WHAT ARE STARS MADE OF? For thousands of years, the question remained unanswered. Until 1925, when the British astronomer Cecilia Payne finished her doctoral thesis. She didn't do this in Britain—although she was allowed to study astronomy at Newnham College, Cambridge, she couldn't graduate because she was a woman. So, after she completed her studies, she went to the Harvard College Observatory to write her dissertation, a piece of work described by her colleagues as "undoubtedly the most brilliant PhD thesis ever written in astronomy."

Payne's specialism was starlight, or rather, she researched what's missing from it. When the light inside a star travels toward the surface, it doesn't emerge into space in one piece; a small amount of it is prevented from passing through by the atoms that make up the star's matter. The electrons in the shell around the atomic nucleus absorb light, and they do it in a very specific way: The wavelengths—i.e., colors of light—absorbed by an atom depend on how many electrons there are and how they are configured. When you separate starlight into its constituent colors and turn it into an artificial rainbow, or "spectrum," certain colors will be missing from that rainbow. The light made up of those wavelengths will have been absorbed by the electrons, and those places will show up as dark lines in the spectrum.

Each chemical element creates its own unique pattern of spectral lines; for instance, hydrogen produces different ones from carbon or oxygen—and that's how the lines in a star's spectrum—i.e., the star's light—can reveal the elements of which it consists.

We already knew this in theory toward the end of the nineteenth century and had been using this knowledge to determine the composition of stars. As Payne discovered during her research, however, we'd been doing it all wrong. For example, spectral lines in the Sun's spectrum indicated, among other things, the presence of carbon, silicon and iron—the same chemical elements that make up Earth. This confirmed our current theory about how celestial bodies are formed: If the Sun and Earth were created at the same time and out of the same cosmic cloud, they should also consist of the same elements. The Sun, we thought, is just like Earth, only much hotter.

Which is true—but not the whole truth, as Payne's research showed. We hadn't made sufficient allowances for so-called ionization: When an electron absorbs energy, it changes its position in the atomic shell—or even abandons the atom entirely, which produces what we call an ionized atom. And when the number of electrons in an atom changes, so does its pattern of spectral lines. So if we really want to know which elements a star contains, we have to look more closely than Payne's colleagues had, and take into account all the different patterns that ionized elements can produce.

Payne did exactly that, and concluded that the Sun does, roughly speaking, consist of the same chemical elements as Earth. But it doesn't resemble Earth in the least, because there are two elements that on their own completely outnumber all other elements in the Sun: hydrogen and helium. The Sun contains a million times more hydrogen than any other element; helium makes up the bulk of the rest, and the remaining elements are present only in tiny numbers.

Stars, then, consist chiefly of hydrogen and helium. Today, thanks to Cecilia Payne (who, when she married, changed her name to Payne-Gaposchkin), this is part of basic astronomical knowledge. When she published her findings, no one was willing to accept that she was right, and her colleagues (all male, of course) even compelled her to describe them as "spurious" in her dissertation. It was only when other scientists increasingly came to the same conclusion as Payne that her opinion was finally taken seriously.

Cecilia Payne-Gaposchkin showed the world how to properly interpret the stars' spectral lines. In her dissertation, she highlighted one particular star: Deneb, in the Cygnus constellation. Not only is it one of the brightest stars in the night sky, but it is also, as Payne puts it, "peculiarly rich in fine sharp lines, many of which are unidentified." She believed that Deneb provides us with a better opportunity than any other star to fully understand stellar spectra. In this, too, she wasn't wrong—since 1943, Deneb's spectrum has been one of the anchor points of the "MK spectral classification system" astronomers still use today.

61

BETA PICTORIS

Visions of an Alien World

BETA PICTORIS IS the second-brightest star in the constellation Pictor ("painter"). Perhaps it's only fitting that it is this particular star that has provided us with the image we'd been looking for so long. The star always had behaved rather conspicuously: At twenty million years old, it's extremely young, and its brightness is still slightly fluctuating. In 1983, it became one of the first stars in whose vicinity astronomers detected evidence of a surrounding disk of dust and gas, and in 1984 they could observe it directly. Over the following years, they found an increasing number of irregularities in the distribution of the dust—there were lumps and gaps, and the whole disk looked oddly "bent." All this indicated that the process taking place in this young star's disk was exactly what they'd expected: Planets were being formed.

And then, in 2008, they actually observed a young planet there. "Observed" is meant literally in this case, because the astronomer Anne-Marie Lagrange and her colleagues at the University of Grenoble were actually able to take a picture of it.

This was new. Until then, the existence of almost every planet we'd found orbiting an alien star had only been established indirectly: They'd been discovered by virtue of the effect their presence has on their environment. For example, planets make the star they orbit wobble slightly, and they can dim its light a little at regular intervals, and that's how we can identify celestial bodies such as these—but to "see" them is usually impossible.

Planets are minuscule when compared to stars, and don't shine of their own accord. The small amount of light they reflect is completely outdone by the stars, which are much brighter, and even when one of our telescopes somehow manages to capture the light reflected by a planet, all we end up with is an image of one more dot among many other dots. Without further analysis, there's no way of telling what the object's mass is, or even whether it orbits the star. That's why we frequently couldn't be sure that what we were observing was actually a planet.

Depending on whom you believe, the first image of an extrasolar planet was taken in 2004, 2005, or 2008. Yet even if Beta Pictoris wasn't the first star whose planet we observed directly, it's nonetheless one of the most spectacular specimens. When trying to take pictures of a planet outside our solar system, astronomers would usually focus on very young stars: Any potential planets in their neighborhood would necessarily be young, too, and therefore very hot; and this heat can be sensed with relative ease using infrared telescopes, which are specifically designed to record the heat radiation emitted by celestial bodies out there in space.

The picture we have obtained of Beta Pictoris is just such an infrared image, showing a planet of about seven times the mass of Jupiter orbiting its star at roughly the same distance as Saturn is from the Sun. We have subsequently observed this planet long enough to clearly see it revolve around Beta Pictoris—the first time in the history of astronomy that we've been able to watch a planet as it circles its star. It'll be 2029 before its orbit is complete, but even so, the picture we have of the Pictor constellation is exceptional. Including some questionable discoveries, the total number of direct observations of extrasolar planets amounts to fewer than two dozen instances. Yet change is coming. The next generation of telescopes both in space and on Earth will find it much easier to capture images like this, and then we'll be able to see such alien worlds regularly. Although there's one thing that won't change quite as quickly: Regardless of how good those instruments are, the images they'll deliver will only show us dots. Extrasolar planets are simply too far away for us to make out their surface in detail. To do that, we'd have to fly over there and take a look for ourselves.

62

72 TAURI

The Star That Made Einstein's Name

THE STAR 72 TAURI is barely visible to the naked eye. It's located in the Taurus constellation, about 420 light-years from Earth, and is a hot, blue star a little more than twice as big as the Sun. At first glance, there's nothing to make it stand out from the multitude of celestial lights.

However, together with twelve other stars, it was the hero of a scientific revolution: 72 Tauri appears as star number ten in Table 1 of the British astronomer Arthur Eddington's paper concerning observations he made during the solar eclipse of May 29, 1919.

In anticipation of the event he organized not one, but two expeditions, one to Sobral in Brazil, the other to São Tomé and Príncipe off the west coast of Africa; he wanted to be prepared for any eventuality, and not risk missing the solar eclipse due to bad weather. Eddington and his team hoped that one—ideally both—places would yield data that would fundamentally change the world of science.

Their objective was to confirm Albert Einstein's general theory of relativity. Einstein had already turned classical physics upside down when he published his special theory of relativity in 1905, and his expansion of that theory, published in 1915, promised an even more wide-ranging revolution. In it, Einstein introduced the world to a universe that can be deformed. Until Einstein, we'd always considered space and time to be immutable; a stage, if you will, on which the laws of physics are acted out. That this isn't the case—that, rather, each of us perceives space and time in our own way, and that how we perceive it changes with the speed at which two given entities are traveling toward each other—is something he'd already demonstrated in his "special" theory. He was now proposing that space isn't a mere backdrop, but something that can be influenced. Every body of mass in the universe, Einstein declared, bends existing space, and the greater the body's mass, the more it bends it; and these distortions in turn affect the motion of every object in space. This means that gravitational force of attraction is in reality only the palpable effect of the curvature of space-time being caused by objects with a large mass: For instance, Earth doesn't revolve around the Sun because it's controlled by a force as if attached to an invisible lead, but because the Sun is bending space around it, and Earth can't help but follow the path of the curve.

It was an astounding and revolutionary statement, but also hard to prove. Nonetheless, this was exactly what Eddington wanted to do. According to Einstein's theory, rays of light also necessarily follow any curvatures in space, so that when a ray emitted by a distant star passes close to the Sun, the curvature in space created by the Sun will bend it a little—with the result that it will slightly shift its position in the sky.

However, the Sun is much too bright for us to be able to observe any of the stars in its immediate vicinity. Unless, that is, you wait for an eclipse. Eddington therefore established which stars would be in the Sun's neighborhood on May 29, 1919—he would observe these stars, determine their apparent positions while the Sun was obscured, and compare those to their actual positions. If Einstein really was right, the positions of the stars would diverge by a specific degree from their known positions.

Observing this wasn't easy; clouds threatened to block the view, and the climate was interfering with the equipment. In the end, though, they managed to capture enough high-quality images of the majority of stars to test Einstein's theory, and it turned out that 72 Tauri had indeed changed position, in exactly the way they'd predicted. The other stars, too, corroborated the general theory of relativity. Eddington's meticulous calculations (which have been checked and validated many times over) demonstrated that bodies of mass distort space. Or, as the man himself put it, "the results [. . .] leave little doubt that a deflection of light takes place in the neighborhood of the Sun and that it is of the amount demanded by Einstein's generalized theory of relativity."

Newspapers all over the world reported on the success of the expedition, and the sensational confirmation of the general theory of relativity. By the grace of 72 Tauri, Albert Einstein became world-famous, and was thenceforth considered the epitome of scientific genius.

63

V1

The Most Important Star in the Universe

WHEN, IN 1923, THE American astronomer Edwin Hubble picked up one of his astronomical images, crossed out an "N" and replaced it with "VAR," he revolutionized our conception of the universe forever. He'd uncovered the secret of the star called V1, and with it the true nature of the cosmos. V1 is the most important star in the universe—at least when it comes to our understanding of its true extent.

"Here is the letter that destroyed my universe," said the astronomer Harlow Shapley when Hubble informed him of his discovery. At the start of the twentieth century, we didn't yet fully comprehend the structure of the universe we live in. We knew that it was big, and that besides the Sun there are many other, distant, stars. But there was something else to be seen throughout the sky: "nebulas," cloudlike objects that don't resemble anything like stars—and scientists were mystified by them.

Shapley and his followers believed that these nebulas were no farther away from us than the stars, and that they were merely the foggy aggregations of the gas between them. According to them, there was quite simply nothing whatsoever in the sky except for the stars we could see. His colleague Heber D. Curtis disagreed, arguing that space was full of "island universes"—stars gathered in large clusters, so-called galaxies—and that the Sun and the stars visible to us in the sky merely constituted one galaxy among many. He contended that nebulas weren't nebulas at all, but remote galaxies separated from the Milky Way by vast regions of empty space.

This altercation, dubbed "the Great Debate," preoccupied the world of astronomy during the first decades of the twentieth century, and it was Edwin Hubble who put an end to it: Using the large Hooker telescope at the Mount Wilson Observatory in California, he was able to make out individual stars in the Andromeda nebula. However, he couldn't tell how far away they were—the technology available in his day was only capable of accurately measuring the distance to the stars that are relatively close to us.

Hubble was looking for novas—i.e., stars erupting near the end of their lives. It was generally assumed that, in this, the stars would always follow the same procedure, and thus also change their brightness in roughly the same way; if he compared a nova's theoretically predicted brightness to its visible brightness, he should therefore be able to estimate the distance between Earth and the Andromeda nebula with a fair degree of precision: The weaker the nova's light, the farther away the nebula had to be.

In a picture taken on October 5, 1923, he spotted three stars that might be potential novas and marked them with an "N." Yet when he compared them to earlier pictures, he noticed that one of the stars was behaving strangely. At times it shone more brightly, at other times more weakly—which was characteristic of a variable star, but not something a nova would do. So Hubble changed its label from "N" to "VAR," and on closer examination realized that the star was in fact a very special kind of variable, a so-called cepheid. A cepheid's brightness fluctuates at regular intervals, and the length of those intervals depends on the intensity of its light. Hubble simply measured the star's pulsation period, then worked out the intensity of the star's light and determined the star's actual brightness. By comparing this to its apparent brightness, he was able to calculate its distance from Earth.

In the course of the following year, Hubble discovered several other cepheids in the Andromeda nebula. The results were always consistent: The Andromeda nebula is located much farther away (by today's calculations 2.4 million light-years away—Hubble thought it was a bit closer) than the maximum possible extent of the Milky Way. The Andromeda nebula therefore couldn't be a nebula, but had to be a gigantic cluster containing billions of stars, which only looks like a cloud because of its remoteness. We now know that the Andromeda nebula is in fact the Andromeda galaxy, and that most of the other supposed nebulas are galaxies, too. At a stroke, Hubble's discovery consigned the Milky Way to being just one among countless islands of stars, in a universe that had suddenly grown far bigger than we could ever have imagined.

Hubble's star V1 gave us our first glimpse of the true size of the cosmos—and to this day we're still trying to fully comprehend and explore the limits of this vast universe.

64

KEPLER-1

Bright Sun, Dark World

WHEN THE KEPLER SPACE TELESCOPE discovered a planet near the star TrES-2, there was much rejoicing—though not for the obvious reason. Scientists had known since 2006 that there was a planet orbiting this star, when it was spotted during the Transatlantic Exoplanet Survey (TrES), but when the Kepler telescope first opened its eyes in space in April 2009, they used TrES-2's planet to check that everything was working as it should. Kepler had been built not to confirm the existence of known planets, but to significantly expand our knowledge of the planets of stars outside the solar system. And it did exactly that. Before it was finally deactivated in November 2018, we discovered 2,662 exoplanets with the help of this revolutionary telescope, more than with any other instrument before it; Kepler tracked down more than half the planets known to us at the time—and it all started with TrES-2.

In the context of the Kepler mission, the star was labeled "KOI-1"—"KOI" stands for "Kepler Object of Interest," and describes stars that we think may be orbited by planets. And because we already knew for a fact about its planet, TrES-2 was also given the standard designation reserved for any star in whose vicinity Kepler would unequivocally prove the existence of one: the name of the telescope, followed by a sequential number. Since TrES-2 was the first such star, it became "Kepler-1."

In order to discover as-yet-unknown planets, Kepler employed the so-called transit method, which involves observing a star's light and measuring its brightness. If a star is being orbited by a planet that passes directly between the star and Earth, the planet blocks out a little of its light—every time it revolves around it, the star dims a little. This periodic change in brightness is small but measurable, and it allows us to establish the planet's existence as well as its characteristics.

For us to observe a planet's transit, however, the planet has to orbit its star on just the right plane. Not all planets do this, which means that if we want to discover lots of them we have to observe as many stars as possible simultaneously. This is why Kepler was dispatched into space: Away from any atmospheric disturbance, pointed always at the same region in the sky, Kepler was able to spend years keeping a constant eye on about 150,000 stars and measuring their brightness.

This is how Kepler managed to discover all those new planets; and in 2011 we realized that even that first look at TrES-2's known planet had been invaluable. The star itself very much resembles the Sun: Its size, mass, temperature and age are nearly the same as our own star's. But the planet they discovered there is highly unusual: It's a little larger than Jupiter (the largest planet in our solar system) but significantly closer to its star than Jupiter is to the Sun. Its orbit around TrES-2 is much tighter than that of Mercury, the planet closest to the Sun. However, the data we received from Kepler also revealed it to be the darkest planet we've ever observed. It reflects less light than a lump of coal. Astronomers found this out by analyzing its phases: When, from Earth's point of view, a planet is located a little to the side of its star—rather than directly in front or behind it—then, depending on its precise position, it receives a little more or less of the light. Just as we can see a full moon or half-moon here on Earth, we can also see a "full planet" or "half-planet." The total amount of brightness measured thus constitutes not only the light of the star itself, but also the light reflected in our direction by the planet.

The effect is minute, but can be measured with a space telescope like Kepler. The amount of light reflected by TrES-2's planet turned out to be incredibly small. Given the planet's great size this can only be explained by its extremely dark color, but we haven't been able to work out yet exactly why it reflects so little light. More space telescopes will have to fix their gaze on it before this celestial body is likely to divulge its secrets.

65

HD 209458

The Star with an Evaporating Planet

THE STAR WITH THE DESIGNATION HD 209458 is situated in the Pegasus constellation. Its light is too feeble for it to be visible to the naked eye, but when you look at it through a telescope you can see its light dimming every three and a half days. This is because it is being orbited by a planet that passes directly between us and the star. This celestial object is unofficially known as Osiris, after the ancient Egyptian god of the underworld, and belongs to a very special class of planets that has confounded us for quite some time.

In astronomical parlance, Osiris is a "hot Jupiter." This is what we call planets that resemble our own gas giant Jupiter in size and composition, but which, unlike Jupiter, are located in their stars' immediate neighborhood. Jupiter takes about twelve years to circle the Sun, reaching an average distance of 485 million miles (780 million kilometers), five times the distance between Earth and the Sun. The law of universal gravitation tells us that the farther a planet is from its star, the longer it takes to orbit it; the planet HD 209458 needs a surprisingly short 3.5 days to orbit its star, so it must be correspondingly close to it—in fact, the two are separated by just 4 million miles (7 million kilometers). Astronomically speaking, this is an absurdly small distance: It's only a tad over four times the star's diameter. By comparison, Mercury, the innermost planet of our solar system, is 36 million miles (58 million kilometers) from the Sun.

It came as a great surprise to us when, at the end of the 1990s, we first observed celestial bodies like this and realized that planets could exist in such close proximity to their star. The second, even greater surprise was that they can be massive, Jupiter-like gas planets. Until then, we'd assumed that planets of that size can only exist at a considerable distance from their star, because during the time the planets were formed those were the only places where there was sufficient material to produce such large celestial bodies—stellar radiation is so intense that it will quickly eject any gas and dust present in the immediate vicinity.

So how was it possible for an object like Osiris to end up there? The answer lies in something called planetary migration, a phenomenon whose significance only became clear to us when we examined extrasolar hot Jupiters. During the first phase of a planetary system, a large amount of gas and dust remains among the young celestial bodies, like rubble scattered across a planetary building site. The gravitational interaction between this material and the planets that are still in the process of being formed can cause the planets' orbits to widen or shrink. Osiris must therefore have been formed far away from its star, and subsequently wandered closer to it.

Unfortunately, its closeness to its sun is doing Osiris no good whatsoever. It's heating up considerably, and practically evaporating. With each second, it is losing up to about 1 billion pounds (roughly half a million metric tons) of its gaseous atmosphere to space. Our observations have shown that the planet is pulling along a tail of gas, just like a comet.

The discovery of hot Jupiters like Osiris has demonstrated to us that, out there, things can look very different from the way they look in our own solar home. There's greater variety among the planets of other stars than anything to be had here—and when we've stopped holding up our own solar system as the measure of all things (as we've been doing for thousands of years), we'll have taken a decisive step toward understanding the universe a little better.

66

PROXIMA CENTAURI
The Star Next Door

AS WE KNOW, the star closest to us here on Earth is the Sun. But even though the Sun is a bona fide star, it plays such a special role in our lives that we don't really think of it as one—to us, stars are the bright spots of light we only see in the sky at night, once the Sun has disappeared behind the horizon. Among those, Proxima Centauri is the one closest to us.

We didn't always know this. Although it's closer to us than any other star, Proxima Centauri is invisible to the naked eye; the Sun has ten times the mass of this red dwarf star, is seven times larger and is 100,000 times brighter. Proxima Centauri can only be seen using an optical instrument, which is why it didn't awaken the British astronomer Robert Innes's interest until 1915. What first attracted the scientist's attention was its motion: When he compared pictures he'd taken of this small spot of light to older images of it, he saw that it had changed position, and done so in exactly the same way as Alpha Centauri.

Alpha Centauri was already a well-known star; after all, it is the fourth-brightest star in the night sky, and easily discernible in our planet's southern hemisphere, in the Centaurus constellation. Back in the nineteenth century, we'd discovered not only that Alpha Centauri is a double star system, but also that, at 4.3 light-years, its two stars were closer to us than any other star we knew of at the time.

And now there was another star, right next to Alpha Centauri; moreover, it seemed to be traveling through the sky in unison with it. This discovery called for further research, and when astronomers measured its distance from Earth they found that this small star is just 4.2 light-years from us, a shade closer even than Alpha Centauri. Innes therefore proposed to call it Proxima (Latin for "nearest").

Alpha Centauri's next-door neighbor hasn't stopped surprising us since. In 2016, we discovered that this small red dwarf is orbited by a planet—not just any planet, but a notable specimen of possibly earthlike conditions. It is a "super-Earth," a terrestrial planet of between 1.5 and 3 times Earth's mass, and extremely close to Proxima Centauri. On the one hand, this is a good thing, because it means that (at least in theory) it receives enough of the star's radiation for its surface temperature to facilitate the existence of life. On the other hand, however, it's also a bad thing, because like many other red dwarfs Proxima is a "flare star": It displays extreme variations in brightness, which also produce high levels of X-ray emissions. Any planet in the immediate vicinity of the star will find itself on the receiving end of the full force of this intense radiation, which is unlikely to promote the evolution of life.

Proxima Centauri's closeness to the Sun and the existence of at least one orbiting planet make it an enticing destination for interstellar missions. All our space probes have so far confined themselves to our solar system, and even the distance to a star as close to us as Proxima is, from a human point of view, still immense and impossible to cover—with today's technology, it would take us several tens of thousands of years to reach it. Some models suggest that we might be able to accelerate miniature space probes using lasers, and perhaps reduce the travel time to a few decades. The probes would be unable to stop there, let alone land, but we'd at least have an opportunity to obtain an image or two of the star and its planet. Seeing an alien planetary system directly and up close for the first time would make it a more than worthwhile effort.

Be that as it may, we have at any rate managed to solve one of Proxima Centauri's riddles: For a long time, we weren't sure whether the star's closeness to Alpha Centauri is merely coincidental, or whether it actually forms part of the system, and in 2016 we finally got our answer: The star orbits Alpha Centauri, once every 600,000 years.

67

NGS

Killed by Laser

"NGS" ISN'T THE NAME of a star, but short for "natural guide star." Theoretically, almost any star can earn the title, as long as it's bright enough—in which case it can help us to circumvent the disruptive influence of Earth's atmosphere, and thus solve a crucial problem for astronomers.

Of course, just like the rest of us mortals, astronomers are in principle grateful for the fact that our planet is enveloped by a life-sustaining atmosphere. Still, it makes things fairly difficult for them: For the last couple of miles of its journey to the mirrors in our telescopes, the light from the stars—which till then has moved undisturbed through space for thousands, even millions of years—has to cross Earth's troublesome atmosphere. As soon as it's no longer moving through the cosmic vacuum but through air, the light is forced to slow down a little. The speed at which the light can proceed through this region depends on the air's density, which in turn is dictated by its temperature. Unfortunately, our atmosphere is continuously in motion: The different layers of air, each with its own temperature and density, are forever swirling through each other, and the resulting turbulence diverts the light from its path.

That's also the reason why we see the stars "twinkle" in the sky. The atmosphere's restlessness shifts their light now this way, now that, and the stars appear to be constantly hopping about in the sky. Consequently, what our eyes perceive as a momentary twinkle is reproduced as a blur by the cameras of our atmospheric telescopes.

We have tried to solve the problem by using so-called adaptive optics: A telescope captures the light of a guide star, which is then directed toward a sensor that measures the degree to which the originally orderly light wave has been distorted. The sensor relays the information to the telescope's mirror, which is then very slightly deformed as the corrective system ensures that the mirror's surface "bends" in exactly the right places to cancel out the distortion in the starlight. In other words, the mirror flickers in the opposite direction to the stars, and thus produces a sharp image.

For this technique to work, the mirror has to be deformed automatically, several times a second. This means that the sensors have to measure, evaluate and communicate the distortions in the starlight in rapid succession, and they can only do that if they can receive enough light from the guide star. If there happens to be a sufficiently bright star like that in the region of sky you want to observe, it can serve as a natural guide star. But these are hard to come by, and in order for adaptive optics to work properly, we've learned to do without natural guide stars and replaced them with laser stars.

Laser stars are artificial stars projected onto the sky using an intense laser beam. They're bright enough to facilitate adaptive optics, and moreover can be placed exactly where you need them. They've been successfully deployed since the 1980s, though initially restricted for use as part of a US military project. However, the technique was finally made public in the 1990s, and today laser stars are helping astronomers to take a better look at the universe without having to leave Earth's atmosphere.

68

M87*

The Invisible Made Visible

ON APRIL 10, 2019, astronomers gave the M87 galaxy a new star. Not a celestial body, but the typographical symbol for a star—an asterisk. The name M87* describes the object whose image was seen by the public for the very first time on that day. The object was a black hole.

Black holes are by definition invisible. The gravitational forces in action around them are so great, and the space so bent, that it's impossible for light, or indeed anything else, to escape. Black holes are as black as black can be—once something has gotten inside one it can never reach the outside world again.

It's not the same for their surroundings, though. When material disappears into a black hole, it does not go unnoticed: Before it vanishes, it's accelerated by the gravitational pull and swirls around the hole at high speed, forming a disk of gas and dust. The material grows so hot that it begins to glow. Among other things, it starts to emit radio waves—which we had detected at the center of our own galaxy as early as 1932.

In 1974, the British astronomer Robert Brown named this radio source at the center of the Milky Way "Sagittarius A*," after the eponymous constellation that marks the galactic center in the sky. The letter A stands for the first radio source detected there, and the "*" is a symbol used in physics to describe certain atomic states. When an atom absorbs a sufficient amount of energy, it emits radiation—and something of that kind, thought Brown, must be happening at the center of the Milky Way, too, although on a larger scale.

We now know that Sagittarius A* is a supermassive black hole more than four million times heavier than the Sun, whose surroundings radiate a vast amount of radio waves into space. You can find objects like this in the centers of other galaxies, too, including the supergiant galaxy Messier 87 (M87), 54 million light-years from Earth. In April 2017, the Event Horizon telescope was pointed right at its center—this global network of synchronized radio telescopes was watching M87 from various positions around the world, which allowed the collected data to be combined to simulate a telescope a few thousand miles in size. Large enough to see the center of the galaxy in detail.

Months of computer analysis produced a visual representation of the radio waves radiated by the galaxy, which showed a bright ring of intense "radio light" with a dark region at its center: the supermassive black hole's so-called shadow, surrounded by a luminous disk. This was the first time we were able to go beyond merely describing a black hole in terms of theoretical mathematics, or to observe it indirectly through its effect on its environment—we had taken an actual picture of it. At least, we'd captured what can be physically captured of a black hole: its dark, impregnable region.

This image initiated a new era in astronomy. Once again, we have succeeded in making the invisible visible, and can now directly examine something hitherto seen only in artistic impressions and computer simulations. Not only that, but our analysis of those images might one day reveal how these ineffable celestial bodies are formed. The black hole at the center of M87 has 6.5 billion times the mass of the Sun, and there's no star in the universe big enough to create a black hole as gigantic as this when it implodes at the end of its life. Black holes of that size must therefore be produced either by an amalgamation of several smaller black holes, or by some entirely different mechanism we are as yet unable to grasp.

By directly observing black holes, we may also finally be able to broaden the foundations of modern physics. We've known for a long time that no current theory can wholly account for black holes—both the theory of relativity and quantum mechanics are out of their league when faced with these extraordinary phenomena. What we need is a combination of the two, and the observational data that's now become available to us may provide clues as to what such a theory might look like, and how it may be formulated.

Following the example set by Sagittarius A*, the supermassive black hole at the center of M87 was initially called M87*. But an object at the center of an astronomical revolution such as this deserves more than merely a combination of letters and numbers. That's why, shortly after its first photo shoot, it was given the name Powehi. This comes from Hawaiian mythology and means "adorned dark source of fathomless creation" (and they say scientists lack all poetic feeling . . .). While black holes are unlikely to have much to do with creation, they're undoubtedly a "dark source" of new, game-changing information.

69

KIC 11145123

The Roundest Star in the Universe

THE STAR WITH THE DESIGNATION KIC 11145123 is 4,000 light-years from Earth. It's a blue giant star with a radius of about 900,000 miles (1.5 million kilometers), which makes it more than twice as big as the Sun. It is the roundest star we've ever seen.

All stars of course are large, round spheres of hot gas, but they also spin on their axes, which produces a centripetal force that squishes the star slightly. For example, the distance from the Sun's center to its equator is 6 miles (10 kilometers) longer than the distance from the center to one of its poles. Which isn't much, but KIC 11145123 is even rounder. Despite its staggering size, the difference between the two distances on KIC 11145123 amounts to just 2 miles (about 3 kilometers).

Which begs two questions: Why is it like that? And how, for heaven's sake, can we measure something like this? The answer to the first question is "We're not sure"; the answer to the second is "By using asteroseismology." Asteroseismology is a marvelous science—it allows us to explore the insides of stars at a distance. It works a little bit like seismology here on Earth: When tectonic processes make the earth quake, shock waves spread through our planet and are diverted and reflected by the various geological layers; and the speed at which these waves spread depends on the density of Earth's layers of rock and metal. By analyzing these waves, we have obtained a fairly detailed conception of how the inside of our planet is structured. However, the stars are very far away, and aren't solid bodies. We can't install any measuring equipment on their surfaces, which means that we can only observe them at a distance and have to adjust our methods accordingly.

To achieve this, astronomers avail themselves of the fact that a gaseous star is in constant motion—the energy produced inside it heats up the star, causing the gas mass to surge back and forth. This produces density waves that travel to the star's surface, where they are then reflected back toward the center. The entire star oscillates much like a vibrating bell.

We naturally can't actually see it happen, but the oscillation has a direct impact on the shape of the star, which in turn leads to small changes in luminosity. If we observe the fluctuation in a star's brightness for long enough and closely enough, we can work out the way in which it oscillates. From this, we can deduce not only the density and temperature of the material inside it, but also some of the star's other characteristics, including its size. Just as a large bell vibrates differently from a smaller one, stars oscillate differently depending on their diameters.

Moreover, some of a star's oscillation modes can be more pronounced at the equator while others occur more vigorously closer to the poles, and we can work out a star's exact diameter and shape by comparing their respective intensity. Astronomers were able to do this for KIC 11145123 in 2016, after the Kepler space telescope had spent over four years observing changes in its brightness and collecting the necessary data. We now know that the star is rounder than any other. What we don't yet know, however, is why.

One possible reason is the fact that it spins quite slowly. It's three times slower than the Sun, which needs a little under twenty-seven days for a single revolution. Yet even given its low rotational speed, KIC 11145123 still ought to be displaying a more noticeable flattening than the one we've both calculated and observed. It could have something to do with its magnetic field: The gas inside a star is electrically charged, and a star's shape is therefore influenced not only by its rotation, but also by the strength of its magnetic field, which in turn depends on the way in which the charged matter moves about the star—and all this plays a part in determining its possible oscillation modes.

Without asteroseismology we wouldn't stand a chance when it comes to examining the processes occurring inside a star. It's a perfect example of how scientific disciplines can complement one another and inspire one another to depart from well-trodden paths and develop new techniques.

70

THE MORNING STAR
Light-Bringer in Disguise

THE MORNING STAR is not only simultaneously the evening star, but strictly speaking not a star at all. The term "morning star" merely describes the celestial body visible as the brightest object in the sky shortly before dawn or shortly after dusk. The body in question is almost always Venus.

Our neighboring planet circles the Sun within Earth's orbit, which is why we never see it very far away from the Sun; the maximum angle between Earth and their respective positions in the sky is just 47 degrees. When the Sun disappears behind the horizon in the evening, it's soon followed by Venus; and when you see the planet in the morning, it won't be long before it's outshone by bright daylight. Whenever our neighbor is visible in the sky, it's hard to miss: It's shrouded in a thick layer of cloud that reflects more than three quarters of the sunlight that hits it, making it shine brightly in the night sky.

The only things brighter than Venus are the Sun and the Moon, and that's why the "morning star" has, like them, become infused with religious and mythological significance over the centuries. In ancient Greece it was known as Eosphoros ("bringer of dawn") or Phosphoros ("bringer of light"), after the son of Eos, the goddess of dawn, and Astraios, the god of dusk. The name Lucifer—denoting the fallen angel, God's adversary—is a direct Latin translation of "bringer of light," and most likely derives from Roman mythology; even the Bible associates it with dawn: "How art thou fallen from heaven, O Lucifer, son of the morning!" (Isaiah 14:12). In Christianity, Jesus is also often described as the "morning star." However, even in those days the astronomical reality behind the mythology was well known. The Greek mathematician Pythagoras of Samos (familiar to every schoolchild for his theorem concerning triangles) was supposedly the first person to realize that the morning and evening star are one and the same celestial body.

Whether you see Venus in the morning or in the evening depends on where in its celestial orbit it happens to be. When it's west of the Sun, it is the first to rise, and we can then see it in its guise of morning star for half a year. After that it starts drawing closer to the Sun, until we can no longer see it from Earth, before vanishing completely behind the star for about three months. It eventually reappears east of the Sun, this time in the evening sky, again for about half a year, then becomes invisible again as it passes between us and the Sun. It repeats this cycle roughly every nineteen months.

These days, people no longer associate Venus's appearance in the sky—be it in morning suit or evening dress—with mythical gods. Yet it causes more confusion than you'd expect: Whenever Venus, after one of its long absences, reprises its role as "bringer of light" in the evening or morning sky, the world's observatories are inundated with calls from excited people reporting UFO sightings.

71

OGLE-2003-BLG-235/MOA-2003-BLG-53

Starry Spectacles

OGLE-2003-BLG-235/MOA-2003-BLG-53—with this designation for an extremely feeble star in the Sagittarius constellation, astronomy really has overdone things a bit. However, there's actually a very good reason for this unattractive combination of letters and numbers, which has to do with the planet that orbits this star.

In July 2003 it became the first planet we discovered there, by means of a brand-new method. Astronomers ordinarily use telescopes with lenses or mirrors to capture starlight, but in this case the lens they used was a star nearly 20,000 light-years away.

The path had been long and tortuous. Nearly a hundred years earlier Albert Einstein had published his general theory of relativity, describing how space can be distorted by the presence of massive objects, and how the motion of each object follows this curvature of space, which we perceive as gravitational force. But light, too, is influenced by it; when a beam of light passes close to a massive object, it ends up being diverted by the curvature in space. Einstein's theory was confirmed in rather spectacular fashion when astronomers measured the positions of certain stars during the 1919 solar eclipse, which provided them with an entirely new method of observation.

The lenses and mirrors inside telescopes serve to redirect rays of light in particular directions. Any body of mass in space can do the same, and act as a "gravitational lens" to reveal what would otherwise remain invisible. When we observe a remote star, for example, all we see is the portion of light it radiates straight at us; if a second star (the lens) passes between us and the first star (the source), the curvature in space that the source light has to traverse changes. Some of the rays are diverted by the lens, and subsequently radiate toward Earth together with the star's normal light. For a time, the source becomes brighter than usual, before gradually returning to normal once the lens-star has moved far away enough again.

These celestial "gravitational lens events" are frequent occurrences, and often diffract the light of entire galaxies to such an extent that their image is doubled, and sometimes multiplied; the presence of a gravitational lens can even cause some galaxies to be smudged into arcs or rings. However, things get really interesting when the lens in question is a star orbited by a planet: When an object like that transits another star, the brightness of the source is increased by the lens, as well as its planet. The effect may only be small, but we can measure it using a good telescope, and subsequently work out the characteristics of the lens-star's planet from the fluctuation in the source light.

There's just one problem with this method: A gravitational lens event like that is always unique. The two stars involved in it have nothing to do with each other—they can be separated by a great distance, and merely appear to be aligned when seen from Earth. As soon as the lens-star has moved far away enough, the event is over and never repeated. But if you can't repeat the observation, it is difficult to confirm the discovery of a planet.

To circumvent this problem, two groups set out to independently search the skies for gravitational lens events: the OGLE (Optical Gravitational Lensing Experiment) project team from Poland, and the MOA (Microlensing Observations in Astrophysics) group, a collaboration between Japan and New Zealand. Both noticed the same change in the same star at the same time, and came to the same conclusion: The lens-star had to be orbited by a planet of about 2.5 times the mass of Jupiter. Since there were two teams involved, the star received not just one designation, but a combination of the respective catalog designations assigned to it by the OGLE and MOA teams—though the name is, admittedly, a bit of a mouthful.

72

ORION SOURCE I

A Rather Salty Star

THE UNIVERSE IS SALTY. At any rate, it's salty in the vicinity of the star called Orion Source I. In the neighborhood of this celestial body located 1,300 light-years away in the Orion constellation, scientists have actually discovered large quantities of sodium chloride crystals. Simple cooking salt, the kind found in any kitchen. But how did it get there, and how is it even possible to find something like that?

They found it thanks to the Atacama Large Millimeter/ Submillimeter Array (ALMA), a radio telescope installed on a 3-mile-high (5,000 meter) plateau in the Chilean desert; and also thanks to the fact that salt can produce radio waves. What's meant here is obviously not a radio program with news bulletins, traffic updates and the biggest hits of the eighties and nineties, but long-wave electromagnetic radiation. After all, normal, visible light is nothing but electromagnetic radiation, and stars can emit their energy into space at all sorts of wavelengths. Salt crystals may not be stars, but they can be "excited"—which happens when atoms or compounds absorb energy, for example from light emitted by an adjacent star or from particles colliding with each other. At some point this surplus energy is discharged again, in this case in the shape of radiation with a wavelength that falls inside the radio spectrum. When a large number of particles do this simultaneously, they produce so much radiation that it can be measured by telescopes on Earth; and since different combinations of atoms produce different kinds of radiation depending on their chemical makeup, our radio telescopes have revealed that there's salt floating around somewhere out there in space. Not only salt—sugar, too, and alcohol molecules, and many other incredibly complex compounds.

As with Orion Source I, they are produced inside the so-called protoplanetary disk. The term describes the material that's left over when a star is formed out of a large cosmic cloud of gas and dust, which collects in a disk around the young star and eventually becomes the material from which planets emerge. All sorts of atoms whizz around there, chiefly hydrogen and helium—the most abundant elements in the universe—but also diverse other elements, including sodium and chlorine. There is, on the whole, too much room in space for two of those atoms to come together, but alongside atoms protoplanetary disks also contain diminutive ancient dust particles, which formed back when the star was but a twinkle in the cosmic cloud's eye, and individual atoms can attach themselves to these dust particles and bond over time.

This is how a large quantity of sodium chloride (as well as potassium chloride and other types of salt), equivalent to the amount of salt in all Earth's oceans taken together, came to exist in Orion Source I's disk. What we don't yet fully understand is how exactly this salt came to exist in the first place. Until we discovered this salty star in 2019, we'd assumed that the conditions in a typical protoplanetary disk were unsuitable for the formation of salt particles. Perhaps Orion Source I is simply an extremely atypical case. But perhaps we'll find more young and salty stars one day, which will force us to reconsider what we think we know about protoplanetary disks.

73

LICH

A Dead Star's Phantom Planets

WHEN THE POLISH ASTRONOMER Aleksander Wolszczan encountered three planets orbiting the star Lich in 1992, his colleagues were bemused. They should have been delighted; after all, they'd been looking for extrasolar planets—that is, planets that orbit stars other than the Sun—for a very long time, and now, at last, they'd found some. However, Lich is an uncommon star, and they were doubtful that the bodies orbiting it could really be described as "planets."

The word *lich* is Old English for "corpse," and is really the perfect name for this celestial body. It's a dead star, a so-called pulsar. Once upon a time, a little less than 2,300 light-years from Earth, there lived a large star of much greater mass than the Sun. When the fuel that powered its nuclear fusion process ran out, it collapsed under its own weight. It turned into a neutron star, which is now compressing roughly the same amount of mass as the Sun into a space just 6 to 12 miles (10 to 20 kilometers) across. A neutron star spins very quickly and regularly, and some emit electromagnetic waves into space at regular intervals. Their magnetic field is churned up by their rapid rotation and ends up sending radiation out into space in the form of two "beacons" of light, just like a lighthouse—though we can only perceive them using radio telescopes. The regularity of the star's rotation results in a very regular "winking."

Yet Aleksander Wolszczan noticed that Lich wasn't pulsating as regularly as it should. There were small divergences, such as could be caused by the presence of other celestial objects in the area making the neutron star wobble a little with their gravitational force. A close analysis of the signals suggested that three objects were involved in this case: one of them a little lighter than Earth, the other two of about four times our planet's mass. Lich was evidently being orbited by three planets.

Only—there shouldn't really be any planets there. When a star collapses into a tiny pulsar, it happens with a mighty explosion. During such a supernova, any celestial bodies orbiting the star are destroyed, and it was entirely unclear where a "corpse" like Lich could have gotten hold of new planets. To this day, the matter remains unresolved. One possible explanation might be that the supernova left behind enough rubble, gas and dust to allow a second lot of planets to be formed. They, too, would then be a kind of "ghost" of the now defunct planetary system. This idea was seized upon by members of the South Tyrol Planetarium, and when the International Astronomical Union launched a competition for the naming of alien planets and their stars in 2015, they called the star Lich and named its three planets after wraiths and undead creatures— Draugr, Poltergeist and Phobetor.

Most astronomers now agree that Aleksander Wolszczan's find does not in fact constitute the first observation of extrasolar planets, if for no other reason than that Lich isn't a proper star, but merely the last remains left over after the death of a star. In the decades since, we have discovered just two other pulsars orbited by "ghost planets," which suggests that the emergence of celestial bodies such as these is a very rare event indeed.

Nevertheless, we didn't have to wait long after Wolszczan's discovery to track down actual extrasolar planets. Three years later we found a real planet orbiting the Sun-like 51 Pegasi, and we've added thousands more such objects to the list since then.

Yet astronomers haven't given up exploring the ghost planets of dead stars, for, even if they can't tell us much about ordinary solar systems, they can teach us a few things about how stars and planets die.

74

SO-102

The Star Gazing into the Abyss

SO-102 IS a cosmic Usain Bolt. It hurtles around the center of our Milky Way faster than any other star, and has done much to augment our understanding of the astonishing creature that dwells there: At the heart of the Milky Way, there's a supermassive black hole more than 4 million times heavier than the Sun.

We'd long surmised the existence of these gigantic objects but were only able to confirm our suspicions beyond any doubt by closely observing the stars adjacent to the center of our galaxy. Imagine our Milky Way as a disk, with a large sphere at its center: Inside the disk the stars, including the Sun, are arranged in spiral arms, and the center is located in the so-called bulge, a ball-shaped region about 10,000 light-years wide. There are many more stars there than in the spiral arms, and they're more closely crowded together. And right in the middle of all these stars is the central supermassive black hole.

The star called S0-102 is there, too, revolving around the center in an astonishingly short amount of time in unison with a group of other stars. The Sun needs about 220 million years to move once through the whole Milky Way, but S0-102, as the American astronomer Andrea Ghez and her colleagues discovered in 2012, does one lap of the central black hole in just 11.5 years. This makes it faster than any other star we know of in the region—and thus a priceless source of information.

A star such as S0-102—unlike the black hole itself, which we call "black" precisely because nothing, not even light, can escape from it—is reasonably easy to observe, and we can deduce some of the black hole's characteristics from its motion. It's because of stars like this that we know for certain that there's a black hole in the middle of our galaxy.

The closer a star is to the object it revolves around, the greater its speed. The same laws operate here as govern planetary orbits around a star, and in the same way that we can work out the weight of a star from the motion of its orbiting planet, we can determine from the orbit of stars like S0-102 the probable weight of the object they're revolving around. At the end of the 1990s, we'd observed other stars flying similarly close around the galactic center and worked out that the object there had to be extremely massive; the radii of their orbits had also given us its maximum possible extent. We therefore knew that there was so much mass packed into so little space at the center of the Milky Way that only a black hole could account for it. And once S0-102, too, had completed its path around the galactic center and we'd analyzed the data, there was no room for doubt. Furthermore, having the observational data of two complete orbits available, Ghez and her colleagues were able to calculate that the black hole has a mass of 4.1 Suns.

At the center of the Milky Way, then, there's an unbelievably massive black hole—this much is clear. These days, we're convinced that these objects exist at the center of all large galaxies, but we don't yet understand exactly how black holes of such monumental mass come into being in the first place. All we know is that they aren't produced by collapsing stars like other "normal" black holes, because there are no stars gargantuan enough.

S0-102 and its fellow stars can therefore rest assured that they'll continue to receive a lot of attention from astronomers, and they'll doubtless help us to uncover this or that other secret some day. Just as long as they don't get too close to the black hole and are devoured.

GRB 010119

Quantum Gravity and Planck Stars

ASTRONOMY HAS A SIGNIFICANT PROBLEM: The theories that most successfully explain the universe don't fit together. There's the general theory of relativity, which we use to describe large cosmic structures, and which explains the evolution of the entire universe from the Big Bang to today; and then there is quantum mechanics, which describes the behavior of the fundamental building blocks of matter. Both are excellent theories and have been confirmed time and again by countless experiments. However, we know that they have their limits. Quantum mechanics can't explain gravitation, and the theory of relativity cannot be applied to the quantum world of elementary particles. What's worse, there are certain phenomena that necessarily require us to factor in both theories—black holes, for instance, or the initial phase of the universe after the Big Bang. These concern massive objects, which are the purview of the theory of relativity, but at the same time involve processes that occur in unimaginably small spaces, and thus fall under the aegis of quantum mechanics.

Neither of these great theories of physics can fully account for those phenomena on its own, and despite laboring for decades we've so far failed to consolidate these two ways of describing reality. A "theory of quantum gravitation" has yet to be found.

There's obviously no shortage of ideas and hypotheses about what such a theory might look like. Perhaps space and time are not continuous phenomena, as the theory of relativity claims; perhaps there are irreducible "blocks of space," tiny building blocks of which all space consists—and if the size of space cannot be infinitely reduced, there may also be such a thing as a shortest-possible unit of time. This, at least, is what some theoretical physicists believe. They back the concept of "loop quantum gravity," according to which the universe consists of webs of lines and knots; rather than being continuous, space is made up of loops woven into a network, much like a piece of fabric. If this is true, it would be impossible for anything smaller than these basic units of space to exist.

Their approach might help to solve the problem of black holes. A black hole forms when a large star collapses under its own weight at the end of its life. If the star is massive enough, no known force is capable of stopping the process. To our knowledge, the collapse results in an infinitesimally small thing that now contains the star's entire original mass. However, this kind of "singularity" cannot exist in the real world. In the actual universe, objects don't shrink to an infinitely small size. To comprehend the phenomenon, we need a new theory, and the theory of loop quantum gravity proposes that such a star may indeed become extremely powerfully compressed, but that the compression will end somewhere, because nothing can be smaller than the fundamental building blocks of space. Thus the structure of the cosmos as such exerts an unstoppable force, which prevents the star's collapse.

The resulting object is called a "Planck star," named after Max Planck, the father of quantum mechanics. We don't know whether Planck stars actually exist, but if they do, they inevitably have to dematerialize again: Instead of turning into a singularity, the collapsing star bounces off the smallest units of space, an event that might manifest itself as a brief but violent explosion.

We have actually observed such phenomena: The universe is repeatedly illuminated by so-called gamma-ray bursts, extremely intense light sources that irradiate the entire cosmos. Most are caused by normal stars ending their lives in an explosion or colliding with each other, but the shortest gamma-ray bursts could conceivably originate from Planck stars.

GRB 010119 is one of these very short gamma-ray bursts. On January 19, 2001, it lit up the skies for just 0.2 seconds. Despite an intensive search, however, astronomers found no afterglow; the place where the flash had appeared remained dark and empty. This is exactly how we would expect a Planck star to behave, but it is still far from proof that these extraordinary celestial bodies exist. Either of our two conventional theories would explain the phenomenon just as well. The search for quantum gravity will therefore doubtless continue—we urgently need an answer to this overarching astronomical question.

76

SCHOLZ'S STAR

A Near Miss in the Stone Age

THERE ARE MORE STARS in the universe than we could ever imagine. However, there is even more nothingness. There's a vast amount of space between the stars—so much of it that a collision between two of them is nigh on impossible.

This means that there's no danger, either, of the Sun ever being obliterated in a collision. So far, the closest it came to it was when Scholz's Star flew past us 70,000 years ago. In truth, it's a slight exaggeration to describe the event as a near miss, given that the closest Scholz's Star ever came to the Sun was a not inconsiderable 52,000 astronomical units. In other words, the other star was 52,000 times farther away from the Sun than our own 0.8 light years—a more than sufficient clearance, then.

Even though the distance to Scholz's Star comprised less than a fifth of today's distance between the Sun and its closest neighbor, Proxima Centauri, our Stone Age ancestors are unlikely to have taken much notice. Scholz's Star is a tiny red dwarf star less than a tenth of the size of the Sun in the Monoceros ("unicorn") constellation. It's so dim that you can't see it except through a large telescope, and astronomers remained oblivious to its existence until 2013, when Ralf-Dieter Scholz of the Leibniz Institute for Astrophysics in Potsdam noticed that the star displayed as good as no sideways movement in the sky. This normally tells us that an object is traveling directly toward or away from us, and when we investigated its motion more closely we saw that it was moving away from the Sun in a straight line. This meant that it had been even closer to us 70,000 years ago.

The star became famous overnight and was nicknamed "Scholz's Star"—a clear improvement on its official catalog designation of WISE J072003.20-084651.2.

Red dwarfs like Scholz's Star are prone to sudden fits of brightness. They're so small that all their matter is in constant motion. Hot gas inside a star's core rises to the surface and then descends back into the core, and this movement—much less pronounced in large stars—creates high-intensity magnetic fields. Magnetic fields discharge energy in a mighty explosion that increases the star's brightness, making it theoretically possible for Scholz's Star to become visible for a few minutes or hours.

We can only speculate as to whether this ever happened, and whether the humans who were around 70,000 years ago caught sight of it. Whatever the case may be, historically speaking this near-collision between the Sun and Scholz's Star left no lasting impression; astronomically speaking, however, it may have: It's quite possible that, all those years ago, the star dislodged some of the numerous comets inhabiting the farthest reaches of the solar system, whose revised orbits might now bring them closer to Earth, and one day even make them crash into us. Still, it's highly unlikely, and it will at any rate take those comets a few million years to make their way to us from the solar system's hinterlands. The universe is damned big—a trite observation, but true.

77

ICARUS

The Light of the Most Distant Star

FROM EARTH'S POINT OF VIEW, Icarus looks unusual. When its light first encountered the mirror of the Hubble space telescope in 2013 it had a lengthy journey behind it, having departed from Icarus an almost inconceivable 9.34 billion years ago, when the universe was less than half as old as it is today.

Never before had we seen such a remote star. The light of any one star is simply much too feeble for us to observe across vast distances, and if it weren't for a cosmic coincidence we wouldn't have seen it at all.

The remote star Icarus (whose official designation is MACS J1149 Lensed Star 1) belongs to an equally remote galaxy. Between us and its galaxy lies a profusion of empty space, as well as a profusion of all sorts of objects, including the MACS J1149+2223 galaxy cluster, which had been the original target of observation. Astronomers had wanted to take a closer look at a supernova in the area, but the images they captured suddenly showed another bright spot of light, which grew brighter during the subsequent years. They analyzed its light and identified it as a hot blue giant—which should actually have been invisible.

The fact that it was is down to the galaxy cluster's mass. As Albert Einstein explains in his theory of relativity, mass bends space, and this curvature affects and diverts light— the galaxy was thus acting as a giant lens, magnifying the light of the objects beyond it. However, this wasn't enough: A "gravitational lens" such as MACS J1149+2223 would merely be able to make the star on the other side shine about fifty times brighter, which isn't enough to make an object like Icarus visible here on Earth.

This is where the coincidence came in. A star in one of the galaxies of the cluster happened to be passing exactly between us and Icarus, and just as you can make an optical telescope more powerful by combining several lenses or mirrors, you can do the same with lenses fashioned out of space-time. In this case, the galaxy cluster's immense gravitational lens joined forces with the single star's "microlensing" effect to boost Icarus's light more than 2,000 times—enough for the Hubble space telescope to pick it up at long last.

In the future, we are going to conduct a targeted search for these special gravitational lenses, which can provide us with information not only about the source of the light they magnify, but also about the celestial bodies responsible for the lensing effect themselves. This will enable us to find out how the stars of remote galaxies are constituted, and whether there are any black holes in the region that can also act as gravitational lenses.

Perhaps then we'll discover how many black holes there were in the young universe, which is an important piece of the puzzle in our search for a better understanding of what happened after the Big Bang. Only very massive stars can become black holes, and how many of these there are in turn depends on how matter was distributed after the universe was born. Being able to see distant galaxies with the help of gravitational lenses thus grants us an insight into the cosmos shortly after its formation.

A blue giant has a life span of just a few hundred million years, and Icarus had therefore shuffled off its mortal coil long before its light arrived on Earth. Its legacy, though, will keep us busy for years to come.

78

SIRIUS
The Flood-Bringer's Dawn

"HELIACAL RISE" IS A PHRASE you don't often hear in everyday life. However, it is one that has been exceedingly useful to astronomy for thousands of years. Four millennia ago, no less than the fate of an entire civilization depended on Sirius's heliacal rise.

Sirius commands our skies to this day. No other star except the Sun shines as brightly. It's larger, hotter and more massive than the Sun and a short 8.6 light-years away from Earth, which means that you can hardly miss it—which was just as well for the ancient Egyptians.

Farmers in the Nile Delta used to rely heavily on the Nile bursting its banks once a year and inundating the surrounding lands. Without this flood, there would be no mud and water to nourish the fields. The Egyptians could tell from Sirius when the annual flood was due: The flooding of the Nile was triggered each spring by the arrival of the monsoon in Ethiopia, and that's when they would see Sirius's heliacal rise—i.e., its appearance in the sky at dawn (hence "heliacal," from the Latin for "relating to the Sun"). The stars rise and set, just like the Sun and the Moon—they don't really, of course, but as Earth rotates on its axis it looks to us as if the celestial bodies emerge from behind the horizon, travel across the sky, and then vanish again.

Yet Earth revolves around the Sun as well as its own axis, so we need to differentiate between Earth's rotation in relation to the distant stars and in relation to the nearby Sun. For example, we can measure how long it takes for a specific point on Earth's surface to regain exactly the same position in relation to the starry skies after one rotation—it takes twenty-three hours, fifty-six minutes and four seconds, a so-called sidereal day (from the Latin *sidus*, "star"). At the end of a "star day," however, we don't see the Sun in the same place as we did at the start, because in the interval Earth has moved another notch along in its orbit of the Sun. We therefore see the Sun from a slightly different angle and have to wait awhile before Earth's rotation around its axis has made up the difference. This takes exactly three minutes and fifty-six seconds, and only when the full twenty-four hours are over is the "Sun day" also complete.

In other words, every day the stars rise another four minutes earlier than the Sun. When the Sun and Sirius, for example, emerge simultaneously from behind the horizon on a given day, we call this Sirius's "heliacal rise"; the next morning, Sirius will appear in the sky four minutes earlier than the Sun, another four minutes earlier the day after, and so on.

When a star is too close to the Sun in the sky, we obviously can't see it. But a few days after its heliacal rise, we can make it out quite clearly at dawn, especially if it's as bright as Sirius. Coincidentally—and fortuitously, for the ancient Egyptians living 4,000 years ago—in Sirius's case this would happen precisely when the Nile's annual flood was in the offing. Its heliacal rise thus heralded the start of the new agricultural season, and every year on that day they celebrated the most important holiday in the Egyptian calendar, the Sothic festival, to welcome the new year.

So it isn't any wonder that we humans have studied the stars for thousands of years. When you knew what was going on in the sky, you could plan and manage things better here on Earth—understanding the starry sky was of religious importance and practical necessity as much as it was an instrument of power.

Now that there are dams to prevent the Nile from flooding, Sirius has lost its significance for Egyptian agriculture. Moreover, due to the oscillation of Earth's axis, the apparent positions of the stars in the sky have shifted in the course of the past millennia, and Sirius's heliacal rise would thus in any case no longer be a useful signal. Today, almost the only people interested in the stars' heliacal rise are amateur astronomers, who observe the sky purely for the joy of the thing rather than for any agricultural concerns.

79

V1364 CYGNI

In Search of Dark Matter

IN THE CYGNUS CONSTELLATION there are plenty of bright stars to attract our attention with their luminosity. But it's well worth taking a look at the feebler stars, too, for they can show us things that can't be seen.

The star with the designation V1364 Cygni is one of these numerous unremarkable stars you can perceive only through a telescope. However, there is one thing that makes it stand out from the rest: It is a "cepheid"—i.e., a star whose brightness fluctuates in a very specific way, which allows us to deduce its exact distance. It can thus help us to solve a puzzle that has stumped astronomers for nearly a hundred years.

In 1933, the astronomer Fritz Zwicky published his observations of galaxies in a galaxy cluster. Just as the planets in our solar system are shackled to the Sun by gravitational force of attraction, galaxies in a galaxy cluster also exert gravitational forces on each other. How strong this force is determines how and in which direction they move. For example, the Sun's gravitational pull is powerful enough for Earth to be forced to remain in its orbit instead of flying off into space. Galaxies arranged in a cluster have to be similarly attached to each other; otherwise there'd be no cluster for us to see.

Yet Zwicky realized something quite extraordinary: The galaxies were moving too quickly. The speeds he measured there would ordinarily have caused them to scatter in every direction in space, because the cluster's gravitational force would be far too weak to keep them in check. He studied the light of the galaxies and used it to work out how many stars each galaxy likely contained; from this, he calculated the mass of the galaxies and also how strong the gravitational force is with which they can affect each other. He concluded that the force was simply not strong enough to sustain the cluster.

As irrefutable as this finding was, so was the fact that there had to be *something* keeping it all together. Zwicky believed that there had to be some additional matter within the galaxies in the cluster, perhaps also between them. Invisible matter. "Dark matter," as he called it.

A few decades later, the American astronomer Vera Rubin encountered the same problem. She wasn't observing galaxies in galaxy clusters, but individual stars inside galaxies. Their movement, too, is influenced by the gravitational forces of other stars. Most of the stars in a galaxy, however, are located at its center; the farther away a star is from the center of its galaxy, the more slowly it ought to be moving, in the same way that the planets in our solar system move more slowly the greater the distance from their orbits to the Sun. But Rubin discovered that the speed of the stars didn't decrease as the distance between them and the center of the galaxy increased. It changed only very, very slightly, as if the galaxy's visible mass barely had any influence on their motion. As if there were a lot more mass there than we could see.

Since then, this phenomenon has been repeatedly verified, including in our own galaxy. In 2018, astronomers used the star V1364 Cygni and a few hundred other cepheids to establish more accurately than ever before the correlation between distance from the galactic center and speed in the case of the Milky Way. Here, too, the amount of visible matter didn't tally with the amount of matter necessary to produce the observed strength of the gravitational force.

This means that there has to be a form of matter in our universe distinct from "normal" matter. It has to be dark—that is, neither emitting nor reflecting light or any other electromagnetic radiation. It's essentially invisible matter, whose existence we can only determine in a roundabout way, through the gravitational force it exerts on its surroundings. Moreover, there has to be as much as five times more dark matter than normal matter for the stars and galaxies to move as they do.

We don't yet know what sort of matter it is. Over the years, all kinds of hypotheses have been proposed, but so far we've been unable to verify a single one of them. For now, all we have is this strange motion of celestial bodies. By studying a sufficient number of them, however, we're sooner or later bound to have our answer.

80

KIC 8462852

The Rise and Fall of an Alien Civilization

AN ALIEN CIVILIZATION is in the process of erecting a titanic structure 1,470 light-years from Earth. It almost completely envelops a star and has been designed to extract an unbelievable amount of energy. At least, this is what the media told us in October 2015.

"We'd never seen anything like this," commented the astronomer Tabetha Boyajian. She and her colleagues had been examining the star KIC 8462852, in the Cygnus constellation, which is almost 1.5 times the size of the Sun and was one of the many objects the Kepler space telescope was scanning for planets. Only rarely is it possible for us to obtain a direct view of the planets of other stars; we mostly have to approach the matter indirectly, and that's why Kepler was keeping its eyes peeled for minute fluctuations in the brightness of the stars. When such a change recurs periodically it means that a planet is transiting the star and blotting out a little of its light each time it passes between the star and Earth.

Although they discovered fluctuations in KIC 8462852's brightness, these were anything but periodic—and sometimes so considerable that it was unlikely that a single planet was responsible for them. On February 28, 2013, for instance, the star's brightness decreased by 22 percent, and didn't return to its original level until two days later. Some occultations were shorter, others longer, some more marked, others less so. Sometimes the star would shine normally for months, at other times its brightness would decrease sharply several times in the space of a few weeks. There was no recognizable pattern, and astronomers were baffled.

No known astronomical phenomenon alone was capable of providing a satisfactory explanation for every aspect of this odd behavior, which is why the astronomer Jason Wright, who wasn't involved in the project at the time, suggested alien activity during an interview about KIC 8462852. He said that the changes in brightness could at least in theory be caused by "megastructures."

Wright did admit that alien activity should only be considered as a last resort in such cases. But KIC 8462852's behavior, he said, tallied rather well with what you would expect if someone were building a "Dyson sphere" out there.

The concept was first described in 1960 by the American physicist Freeman Dyson. Although the Sun radiates a huge amount of energy into space, we can only utilize a fraction of it. Dyson wondered: What if we could construct a shell around the Sun? It would have to be a gigantic structure, a hollow sphere the width of Earth's orbit, with the Sun at its center.

In practice, building such a colossal structure poses all kinds of challenges. You would have to strip whole planets to obtain all the necessary material, and would have to devise entirely new methods to turn it into a sphere stable enough to enclose the Sun. And even if we succeeded, plenty more awkward problems would remain to be solved. It would be easier to build a "Dyson swarm," a large number of energy-collecting structures that swarm around a star, gathering up whatever energy they can. Instead of a single spherical shell surrounding the entire Sun, we could build countless giant platforms, which would store up the Sun's energy as they circle it on independent orbits.

A Dyson swarm under construction could explain KIC 8462852's fluctuations in brightness, but it's unlikely that we'll actually see any signs of an alien stellar building site there. Subsequent observations of the star revealed that its light isn't being blocked out completely, but that it becomes darker in some wavelengths of light than others. This precludes the possibility that a solid object such as a planet—or, well, an alien structure—is obscuring it. But it does tally with what large dust clouds do when they cover up starlight.

Where this dust might have come from is unclear. Perhaps it came from a comet or was flung into the star's neighborhood during a planetary collision. Then again, the star might be changing its brightness of its own accord. KIC 8462852 remains an astronomical conundrum, but we probably won't find any aliens there.

81

STAR 23
The Riddle of the Sky Disk

STAR 23 MEASURES less than a third of an inch (about a centimeter) across. It's fashioned of gold, 4,000 years old and man-made. We didn't discover it in the sky, but in the soil in central Germany, buried near Nebra in Saxony-Anhalt. Star 23 is part of the famous Nebra Sky Disk, and despite its earthly origins not a mite less interesting than its cosmic models.

When the sky disk, which had spent a very long time in the dark, was rather brutally dragged into the outside world in July 1999 by looters in the Ziegelroda forest, no one understood just how impressive an object it is. This Bronze Age find is the oldest known physical representation of the sky; it has given us an insight into the conceptual world of the people who lived back then, and utterly changed our image of prehistoric central Europe.

The disk has a diameter of just over a foot (roughly 32 centimeters). On its iridescent green surface (which used to be a dark, black-blue color), the Sun, Moon and stars are arrayed in gleaming gold. Of the original thirty-two stars on the disk, only thirty remain. Seven of them are assembled in a group we immediately recognize as the Seven Sisters, the Pleiades, which are clearly visible in the sky even to the naked eye. A large golden circle suggests the Sun or the full moon, and next to it, just as large, there's a crescent moon.

Detailed analysis has revealed that the disk contains encoded astronomical knowledge. It's a model of the actual sky, and if you set the two side by side you can tell whether you need to add a leap month to reconcile the lunar and solar years: The length of a year differs depending on whether your calendar is based on the path of the Sun or on the Moon's motion, and if you want to synchronize them, you have to insert leap days at specific intervals. The sky disk shows you when to do this, because when the lunar and solar years have diverged too far, the Moon and the Pleiades will be aligned in the sky exactly as they are on the disk.

Yet the disk is far more than merely a Bronze-Age reminder to adjust your calendar. Even at the time, it was a unique object of great value, symbolizing the relationship between our earthly life and the heavenly dealings of the gods. It demonstrated power over the cosmos, as well as power over the precious metals of which it was created. It was a visible sign of authority and legitimized its owners' mastery over the people.

The sky disk has taught us that complex Bronze Age cultures existed not only in Egypt and Mesopotamia, but also in central Europe. Of course, at first many claimed that it was a forgery—not least the looters and fences, who in that case would have been dealt shorter sentences. Among other things, however, Star 23 proves that the disk has to be authentic. The sky disk was modified multiple times during its existence: Originally, it displayed only the stars, the Sun and the Moon; later, golden arcs were added to the edges, carrying information about the times of sunrise and sunset in central Germany. Chemical analysis has shown that the gold used for the symbols representing the celestial bodies came from the same source. Except for Star 23, that is, whose chemical composition exactly matches that of the arcs added later. When you look closely at the disk, it's plain to see that the star must have been in the way of one of the arcs, and was consequently moved elsewhere. Metallurgical analysis also revealed that the disk was worked on at other times, too—all in all, a production process so complex as to make it difficult to fake.

The sky disk of Nebra is real—and exquisite proof that the sky has enthralled us humans from the very first.

82

SN 2008ha

The Sky Belongs to Everyone

IT WAS FRIDAY, November 7, 2008. And while most teenagers celebrated the start of the weekend and did what teenagers usually do, Caroline Moore, a fourteen-year-old New Yorker, was otherwise engaged. She was busy looking at astronomical images on her computer. She was a member of the Puckett Observatory Supernova Search team, part of a collaboration of volunteers and members of the public interested in astronomy, who combed through images of stars looking for abnormalities.

The pictures had been taken by robotic telescopes and passed on to amateur astronomers like Caroline. On this day, she'd already been working on the project for seven months and had scanned countless images. But her concentration evidently didn't flag, because that evening she found what she'd been looking for: a supernova, the final burst of light from a star dying in a large explosion.

The two images of the same region of sky that Caroline was looking at had been captured at different times, and were identical except for a single spot of light, which you could see in one image but not the other. She reported her discovery, and so, in addition to the teenager's eyes, astronomical instruments now, too, cast their (distinctly larger) eyes in the direction of the galaxy in the Pegasus constellation nearly 70 million light-years away, where the supernova had taken place.

The discovery was confirmed, the supernova was given the official designation SN 2008ha, and Caroline Moore was awarded the title of the youngest person ever to discover a supernova. Furthermore, it proved to be a truly unusual stellar explosion: much dimmer than you would normally expect from a supernova. That the life of a star could end in a bang that was more of a whimper was news to us. Her discovery thus provided astronomers with an entirely fresh insight into how stars die.

It also exemplified the fact that astronomy is a science unlike many others. You don't have to have spent years studying the subject to make a valuable contribution to it. Everywhere in the world there are people who pursue astronomy as a hobby, who simply enjoy surveying the skies, and organize themselves into clubs, private observatories or collaborations like the one in which Caroline Moore took part. Private stargazers have discovered countless comets, asteroids and supernovas, found the planets of other stars and assisted professional researchers in the evaluation of huge chunks of data. In a research field as vast and sometimes unwieldy as the universe, contributions from members of the public can be so invaluable that there's now a term for it: citizen science.

Unlike in many other sciences, it doesn't take much to conduct astronomical research. All you need is your eyes, a dark sky and the desire to apprehend the universe—or merely to enjoy its beauty. The universe is big enough for us not to run out of research subjects anytime soon, and each of us has a chance to discover something new there. The sky belongs to everyone.

83

SPICA

Climate Change and Celestial Mechanics

SPICA IS THE BRIGHTEST STAR in Virgo. Its name means "ear of wheat" in Latin. Both in Mesopotamia and, later, in Rome, people thought that they could make out the goddess of farming in its constellation. In a way, Spica does have something to do with agriculture—or at least with our understanding of Earth's changing climate.

The regular succession of the seasons on Earth is the result of Earth's axis not being positioned perfectly vertically on the plane on which it orbits the Sun. The planet's rotational axis is tilted by about 23.5 degrees. Thus, at different times of the year, the northern hemisphere of the globe is inclined some-times toward the Sun, sometimes away from it (the same goes for the southern hemisphere, too, of course, only the other way around). When Earth's northern hemisphere is directed toward the Sun, the Sun spends more hours in our sky; its light falls at a steeper angle and can thus deliver more heat for longer (in winter it's the other way around). Only twice a year do the light of day and dark of night last equally long. These dates mark the beginning of the astronomical spring and autumn, respectively, and are called the "vernal equinox" and "autumnal equinox"; the same name is given to the Sun's positions on those days.

You can measure the distance of any given star in relation to these positions. (In fact, the vernal equinox is a reference point for precisely such measurements in modern astronomical coordinate systems.) Regardless of how often you measure a star's distance, the result should always be the same. Yet in the second century BCE, the Greek astronomer Hipparchus discovered that his measurements didn't coincide with those of his predecessors. Spica's position had shifted noticeably in relation to the autumnal equinox. From this he concluded that Earth's rotational axis had changed direction over time.

Today, we know that he was right. Earth's rotational axis doesn't always point in the same direction, but instead describes a circle over the course of roughly 26,000 years. This is due to the gravitational forces of the Sun, the Moon and the other planets in the solar system. This is not without consequences: Currently, the northern hemisphere's winter always occurs when Earth is at its closest to the Sun, but in about 11,000 years Earth's axis is going to point in the opposite direction. Then it'll be summer in the northern hemisphere when Earth is closest to the Sun, and winter when it's farthest away, and the seasons will be a little more marked than they are today—the winters, especially, will last longer.

However, Earth's path around the Sun is also subject to variation. It wobbles around a bit in space, grows larger and smaller, sometimes deviates more from its orbit, sometimes less. All of which means that we receive varying amounts of solar energy. These variations, however, are fairly undramatic: Generally speaking, Earth's orbit is quite stable. But when the changes intensify in a particular way, they produce a kind of feedback that turns Earth much warmer or colder, resulting in ice ages as well as "warm ages," and changing the entire planet's climate. The impact of these changes in Earth's movement on our climate are called Milankovitch cycles, after the Serbian geophysicist Milutin Milankovitch, who studied them at the beginning of the twentieth century.

In the meantime, of course, we have learned that celestial mechanics are not the only reason for climate change. Aside from geological phenomena such as volcano eruptions, today it is we humans who exert the greatest influence on Earth's climate, and the effect we have makes itself felt much more quickly than the Milankovitch cycles, which take place over periods lasting between 10,000 and 100,000 years. Owing to our ever-increasing emission of greenhouse gases, we have raised the CO_2 content of Earth's atmosphere by 45 percent in the past 150 years, heating Earth at a far greater rate than any process of celestial mechanics could. We have to take charge of this problem, and we have to do it now. It's safe to say that the goddess of farming won't be able to help us.

84

FELIS

Celestial Ex-Cat

FELIS IS THE BRIGHTEST STAR in the nonexistent constellation also called Felis (Latin for "cat"). While the sky is packed with dogs—Canis Major, Canis Minor and Canes Venatici ("hunting dogs")—as well as their forebear, Lupus ("wolf")—the ineffable domestic cat is absent from the stellar menagerie.

This evidently bothered the French astronomer Jérôme Lalande. In 1799, while recording a few stars in his sky chart, he decided to outline a new constellation between Hydra and Antlia. He justified his choice of animal by explaining that it was prompted by Claude-Antoine Guyot-Desherbiers's "Poème du Chat": "I love these animals very much. There is a large empty space on the charts, and I have filled it with this image [. . .]; the starry sky has tired me enough in my life, that I shall now have my fun with it." He begged the German astronomer Johann Elert Bode to include the cat in his sky atlas, and thus it appeared in the latter's 1801 *Uranographia*, on plate 19.

Yet the celestial cat was denied a long life. The constellation was nondescript, and its stars shone only timidly. Felis, the brightest among them, is 530 light-years from the Sun, and despite being an orange giant star it is only visible without a telescope in a thoroughly dark sky.

Having made its debut in Bode's atlas, the cat was adopted by various other European star maps (and for a time mislabeled by one of them as a hare). You can follow its tracks on celestial globes and atlases until 1888. Yet the Americans proved rather more reluctant about the whole thing—their reference works entirely ignored Lalande's cat.

When the International Astronomical Union reorganized the sky into its eighty-eight official constellations in 1928, the cat's fate was sealed. Unlike its fluffy relatives on Earth, their celestial counterpart hadn't managed to win enough admirers. It was considered outdated, and excluded from the universally accepted modern list of constellations. Thus, while you can find forty-two animals in the sky these days (including beasts of myth and fable such as the dragon Draco, Pegasus and Phoenix), you'll search in vain for a cat. Felis's stars were simply redistributed among the constellations of Hydra, Antlia and Pyxis. Cat lovers have to content themselves with the constellations of Leo, Leo Minor and Lynx—but there is some reason to celebrate: In 2018, the International Astronomical Union decided to officially confer the name Felis on the star with the catalog designation HD 85951. At least the brightest star of the ex-constellation can keep the memory of the celestial cat and its brief career alive, if only in name.

85

WASP-12

Wet Asphalt in Space

WASP-12 IS A SUN-LIKE STAR in the Auriga ("charioteer") constellation. You can't see the star with the naked eye, but in 2008 the British-led planet hunters WASP (Wide Angle Search for Planets) pointed their telescopes at the star and discovered an orbiting planet. The planet with the designation WASP-12b is slightly heavier and nearly twice the size of Jupiter, the largest planet in our solar system.

This alien celestial body is a dark world. It absorbs 94 percent of the light that meets it and is darker than freshly spread asphalt. Then, in 2013, the Hubble space telescope detected the presence of water vapor in its atmosphere.

Of course, there isn't actually any asphalt there; nor are there any streets, or rain to soak those nonexistent streets. WASP-12b is an utterly barren world. It's a gas giant, which means that it doesn't have a solid exterior, and it's so close to its star that its surface temperature is more than 3,600 degrees Fahrenheit (2,000 degrees Celsius). But precisely because it is so large and so close to its star, we can see things there we wouldn't be able to see anywhere else. The planet itself is invisible to our telescopes; the bright light of its star completely outshines it. But when it transits the star, a small amount of the starlight manages to get through the planet's atmosphere and radiates all the way to us. In the process, some of the light is absorbed by the atoms in WASP-12b's outer layers—each chemical element does this in its own particular way, and by analyzing the light that reaches us we can find out which atoms and molecules exist in WASP-12b's atmosphere.

In this case we found water, among other stuff—but that doesn't prove that the planet is inhabited by living things. There's an abundance of hydrogen and oxygen in the universe, which can combine into water molecules under the right conditions, and we've thus found water not only in the atmospheres of some alien planets such as WASP-12b, but also in the large gas clouds between the stars. Of course, on WASP-12b the water is present in the form of gas—i.e., as water vapor—no other state would be possible on a hot gas planet like that.

Water vapor in a planet's atmosphere, then, isn't necessarily a sign of life. It would be another thing entirely if we discovered pure oxygen somewhere, because oxygen can very easily combine with other chemical elements. The only reason that Earth's atmosphere is made up of 21 percent pure oxygen is because it's inhabited by a plethora of plants, which keep producing oxygen as part of their metabolic processes; if life on Earth were extinguished, the atmosphere's oxygen content would also quickly disappear. There are a few molecules besides oxygen, for instance methane, that clearly point to the existence of life. Were we to find such "biomarkers" in the atmosphere of a planet that isn't a gas giant, but terrestrial, and whose orbit runs neither too close nor too far from its star, this would constitute pretty decent evidence that life isn't merely possible there, but actually exists.

Our telescopes aren't yet powerful enough to examine planets of this kind, except for large and hot planets like WASP-12b. But when, in a few years, the next generation of giant telescopes goes into operation, we can commence our search for inhabitable alien planets in earnest.

86

ULAS J1342+0928

Shining a Light into the Cosmic Dark Age

THE LIGHT FROM ULAS J1342+0928 has taken 13.1 billion years to reach Earth. The object with this complicated name is a so-called quasar, and isn't a star at all. When we first encountered it, we had no idea what it was. It looked like a star, but couldn't actually be one, so we provisionally called the new discovery a "quasi-stellar object," or "quasar" for short.

These objects first came to light in the 1960s, thanks to radio astronomy. Stars, galaxies and other celestial bodies emit energy in not only the form of visible light, but also elsewhere in the electromagnetic spectrum, for example as radio waves. Radio waves are a form of light, but their wavelength is too long for us to see them directly. However, our radio telescopes were detecting a host of very powerful and seemingly dot-shaped radio sources in the sky, and when we used normal telescopes to find out what strange celestial bodies were broadcasting these radio waves, we discovered something unexpected. Ordinarily, radio waves are caused by galaxies, but in this case astronomers saw something else entirely: radio waves without an associated galaxy. They initially explained the curious phenomenon by suggesting that the galaxies were too remote—the objects had to be very far away indeed for their radio waves to have taken billions of years to reach us, and for us to be unable to identify their source with conventional instruments.

It was only later that we realized that the radio telescopes had discovered the luminous centers of distant galaxies, and that the large amount of radiation was being produced by the supermassive black holes located there. At the center of every large galaxy, there's a black hole with a mass millions—even billions—of times greater than the Sun's. And when any matter, for example interstellar gas, gets close to the black hole it's liable to fall into it. As the material speeds up and becomes hotter, it starts to emit energy, some of it in the form of radio waves. Galaxies like our Milky Way are already rather old and have only a little material left at their center that can fall into the hole. But young galaxies are still full of cosmic gas clouds, which is why the black holes at their centers are so bright.

It is these young galaxies that we see when we look deep into space. Their light takes a long time to get to us, and the longer its journey the further into the past we see. Galaxies like ULAS J1342+0928 are among the youngest we've ever encountered, and when we discovered this quasar in 2017 it was the most remote known object of its kind. When its light set off from the neighborhood of the supermassive black hole toward Earth, it had been only 690 million years since the Big Bang happened.

Studying such quasars is a unique opportunity for us to find out more about the universe's childhood, especially about the phase known as the Dark Age. Shortly after it was born, the cosmos was still far too hot for whole atoms to exist—due to the extreme temperatures, the electrons, which normally form the shells of atoms, were moving far too rapidly to attach themselves to atomic nuclei. It wasn't until 380,000 years after the Big Bang that they were able to do this, when the universe had sufficiently cooled down. And only from that point on was light able to spread through space unhindered. Till then, it had been constantly diverted by the electrons whizzing all over the place. The radiation that subsequently spread in all directions can still be observed today in the form of "cosmic background radiation," and for a long time it was the only light in the universe.

At the youthful age of 380,000 years, the cosmos had entered its Dark Age: There was nothing there to prevent the light from diffusing, but also nothing capable of producing more of it. Stars didn't exist yet—they still had to be formed by the enormous gas clouds so rich in atoms.

It took a few hundred million years for stars and galaxies to come into being. Their radiation caused the ionization of some of the atoms—i.e., detached electrons from the atomic shells—but by now the universe had grown large enough for all that new light to pass through it. This phase is known as the epoch of reionization and marks the end of the Dark Age. How exactly the transition happened is something we don't yet fully understand. But the radiation emitted by galaxies like ULAS J1342+0928 originates from that period and will help to cast a little light into the universe's Dark Age.

SANDULEAK -69 202

The Long-Awaited

ON FEBRUARY 24, 1987, our wait was finally over. The telescope of the Las Campanas Observatory in Chile had spotted something never seen since these instruments were invented: a supernova in our immediate vicinity.

The last time this happened was in 1604, before the advent of the first telescope, when astronomers including Johannes Kepler saw a star suddenly conjure itself and its blinding light out of thin air. But we had to wait nearly another 400 years before a nearby star died in a mighty explosion of light. Particularly in the twentieth century, this rather tested astronomers' patience: By then, they not only knew exactly what these events were, but had the necessary technical resources at their disposal with which to closely analyze them, assuming that they didn't occur too far away. They had already observed a number of supernovas in other galaxies, which were bright enough to see across even those vast distances. But what they really needed to conduct a detailed investigation was a stellar explosion in our own Milky Way.

The star that exploded in February 1987 almost fulfilled this requirement. It was just under 170,000 light-years away—technically outside our own galaxy, yet still within the bounds of the Large Magellanic Cloud, one of the dwarf galaxies that accompany the Milky Way like close satellites.

Astronomers weren't quite sure at first which star it was that had thus violently ended its life. They found the answer in a catalog compiled by the Romanian-born American astronomer

Nicholas Sanduleak in 1970: Among the 1,275 stars recorded in that catalog was an entry designated "-69 202," which had till then led but a shadowy existence. No one had been interested in this unremarkable celestial body before, but as it turned out it was this very star that had disappeared. And so Sanduleak -69 202 became an instant celebrity—even though it no longer existed.

Where shortly before a large, hot, blue giant star had led its inconspicuous life, now there was nothing but a couple of rings of gas and dust: the last remains of the stellar atmosphere cast off into space by Sanduleak a few tens of thousands of years before its demise. Those hitherto barely visible rings were now glowing with the heat of the supernova's radiation. Four years later, we were even able to watch as debris from the ruined star collided with the dust rings.

Even more interesting than that were the observations made a few hours before we came across this supernova. Three particle detectors positioned in different locations on Earth had registered a sudden spike in neutrinos—these extremely light and fleeting elementary particles are produced, among other things, during the nuclear reaction that occurs inside a star, and a particularly large number of them during supernovas. They barely interact with other matter, which is why nothing stood in their way as they shot out of Sanduleak's core into space right after the explosion. Meanwhile, the light from the explosion still had to penetrate the star's matter, which kept diverting it from its path, and therefore left the star only some time after the neutrinos; and since neutrinos travel almost at the speed of light, it hadn't managed to catch up with them by the time they reached Earth.

Our observation of those neutrinos, and the detailed data provided by our telescopes, have helped us to refine and expand our modeling of supernovas. But what we really need is an actual stellar explosion inside the Milky Way. We've already missed one of them: It looks like a star exploded in the Sagittarius constellation, just under 27,000 light-years away, at some point between 1890 and 1908. The reason the event escaped our notice is that our view of the site was being obstructed by several cosmic dust and gas clouds. We didn't encounter the supernova's leftovers until 1984.

Patience is a virtue. However, on average one or two stars ought to explode in the Milky Way every century, so another will perish sooner or later. Hopefully this time we'll be in a position to make the most of our front-row seat.

88

3C 58

Star Full of Quarks

ON AUGUST 6, 1181, a guest star appeared in the sky. Chinese astronomers used to bestow this welcoming name upon any star they hadn't seen before. Today we call this phenomenon a "supernova"—a large, old star exploding at the end of its life, sending up a bright flare that can linger in the sky for months.

Alas, there are only few contemporary accounts of this twelfth-century supernova. But given what we now know about them, it's not beyond the realm of possibility that the Chinese and Japanese astronomers had witnessed the birth of a "quark star," nearly a thousand years ago. When a star of sufficient mass dies, it creates a "neutron star" when it finally implodes: As the star becomes more and more compressed by its own weight, the negatively charged electrons and positively charged protons that make up the atoms in its matter are practically being pushed inside each other, which results in neutrally charged neutrons; standing shoulder to shoulder, these now make up the entirety of the dead star's mass.

However, neutrons aren't elementary particles. They consist of so-called quarks, which are generally believed to be the basic building blocks of matter. You need three quarks to make one neutron, two "down quarks" and an "up quark." Under normal conditions, it's quite impossible to deconstruct a neutron (or any other particle made of quarks), because the force holding the quarks together has a singular characteristic: It doesn't weaken when two quarks are separated. To take a neutron apart, you have to expend an enormous amount of energy; and even when you do have enough energy to separate two quarks, that same energy would immediately create further quarks. Moreover, as Albert Einstein has taught us, mass and energy can transform into each other; the separated quarks will therefore instantly merge with the new ones created by the neutron's absorption of that energy to form fresh neutrons, as well as other particles.

Quarks, then, can't exist as independent particles, unless the prevailing conditions are particularly extreme, such as they might be inside a neutron star. Inside a sufficiently massive neutron star, the neutrons may become so powerfully compressed that they can release quarks—and the star would then continue to collapse into itself until it has turned into a "quark star."

Whether a thing like that actually exists is still unclear. Current thinking has it that a too-heavy neutron star will merely keep imploding until a black hole emerges. But if such a thing as "quark matter" exists, so does that peculiar species of star.

For quark matter to form, the mass of the dying star has to be great enough to be capable of compressing its matter beyond the state of a neutron star, yet small enough to prevent the whole thing from completely imploding into a black hole. At first glance, a quark star would look little different from a neutron star, and consequently be difficult to identify. Nevertheless, the remains of the 1181 stellar explosion may well be just such a quark star: At the location where the "guest star" appeared—in the Cassiopeia constellation—you can now see the neutron star with the designation 3C 58. Although we can't be sure that it really is the leftovers of that twelfth-century supernova, this star is nonetheless incredibly interesting, having a characteristic that we believe to be peculiar to quark stars:

It is distinctly cooler than a neutron star has any right to be. This may not constitute absolute proof, but it does suggest that the universe still has a few unexpected curiosities in store for us.

89

CoRoT-7

Home of the Super-Earth

THE STAR CoRoT-7 IS ORBITED by a super-Earth. This doesn't make that planet somehow better than our "normal" Earth—quite the opposite in fact: It has a surface temperature of about a thousand degrees Celsius, and when it rains, it doesn't rain drops of water, but rocks. (This is because the planet's extremely high temperatures cause some of the rocky surface to evaporate; upon reaching the atmosphere it cools, solidifies, and then drops back down to the surface.) "Super-Earths" are, rather, what astronomers call a class of celestial bodies not found anywhere in our solar system.

We can divide the Sun's eight planets into two groups: The four closest to the Sun—Mercury, Venus, Earth and Mars—are rocky spheres with a metallic core and a solid surface; by contrast, the outermost planets, Jupiter, Saturn, Uranus and Neptune, are "gas giants." The latter consist almost exclusively of hydrogen and helium and have no solid surface. Uranus is the smallest of the gas giants and has more than fifteen times the mass of Earth, which in turn is the largest of the rocky celestial bodies, the so-called terrestrial planets.

There is no other kind of planet in our solar system. However, alien stars are sometimes orbited by planets that are distinctly heavier and bigger—planets such as CoRoT-7's aren't gas giants but are nevertheless in the Monoceros constellation. The planet was discovered in 2009 by the French space telescope CoRoT; when astronomers measured the brightness of its Sun-like star in the Monoceros constellation, they found that it was susceptible to the characteristic fluctuations that indicate the presence of a planet. Each time a planet transits its star, the star grows a little darker; in the case of CoRoT-7, this happens every twenty hours. So the planet needs less than a day to revolve around its star. The reason for this is that it's very close to it, which also explains the planet's high temperatures.

CoRoT-7's planet is 1.7 times larger than Earth and has nearly five times its mass. This celestial body therefore has to be a lot denser than Earth. It must be a planet constructed of a very large iron core surrounded by a thick, dense layer of rock. Objects like this—terrestrial planets that are larger and heavier than Earth—are called "super-Earths"; and there seems to be no shortage of them in space.

It's still unclear why there are so many of them around alien stars, but none in our solar system. We also don't know whether life is possible on these planets. The studies published to date have come to the conclusion that super-Earths may experience a lot more geological activity than Earth and might also— because of their large mass and therefore more powerful forces of attraction—have much denser atmospheres than ours. This could make them more effective than Earth at protecting their surfaces, and those who reside there, from cosmic radiation. Although very intense earthquakes and volcanic eruptions tend to discourage life, a planet actually needs active plate tectonics to create a climate conducive to life, because slowly shifting continents and the recycling of rocks regulate the levels of carbon dioxide retained by the rock. Super-Earths, then, really may be better than our own Earth—as long as they don't dwell as close to their star as CoRoT-7's does.

To fully understand super-Earths, we'll have to explore them in greater detail. Yet they've already taught us one thing, if nothing else: Our solar system is anything but typical. This glimpse into space has revealed that the planets of other stars can be unlike anything we've ever known; and it remains astronomy's greatest challenge to seek out phenomena that we can only envision once we've found them.

90

CYGNUS X-1
The Black Hole's Bright Light

THERE'S A BRIGHT BLACK HOLE in the Cygnus constellation. Or rather, it's the surroundings of this black hole that are highly luminous—after all, a black hole is called "black" because nothing can escape it, not even light. However, the brightness of this particular hole has nothing to do with its having an especially great mass; Cygnus X-1 is only fifteen times heavier than the Sun, and many stars in the universe are a great deal heavier than that.

A black hole forms when a large amount of mass is compressed into a very small space, which sometimes happens when a star dies: As the star nears the end of its life and its fuel runs out, it can no longer sustain the nuclear fusion processes inside its core and collapses under its own weight. Certain factors, such as particular types of forces, can stop the collapse, and in those cases the star turns into a white dwarf or a neutron star, but if the celestial body's mass is too great for any force to stop its collapse, the stellar matter implodes under the pressure of its gravitational pull. The object becomes smaller and smaller, and its matter becomes ever more compressed.

Such is the fate of stars that are very massive. These transform into black holes, and the reason that black holes are black is that they have an "event horizon." To permanently extract yourself from the influence of a heavy object, you have to move away from it at a sufficiently high speed—on Earth's surface, this "escape velocity" amounts to about 7 miles (11 kilometers) per second—and anything that moves more slowly than that eventually drops back onto its surface. The greater the force of attraction, the greater the required escape velocity: On the surface of the Sun, for example, the required velocity amounts to 383 miles (617 kilometers) a second. The larger the mass of a celestial body, the stronger its gravitational force—but its gravitational force also increases the closer you get to that mass. The Sun's vast size thus limits its force of attraction: Even if you come within close proximity of its surface, its center is still 430,000 miles (700,000 kilometers) away.

Were you to squeeze the Sun's mass into a smaller volume, you'd be able to get much closer to it; you would then experience a far stronger pull from all that mass, and the required escape velocity would increase. If you compressed the Sun's mass into a sphere with a radius of 2 miles (about 3 kilometers), you'd have to travel faster than the speed of light to detach yourself from its surface. However, since nothing can move faster than light, it would be impossible for you to permanently break free from this sphere. The Sun would effectively be a black hole—and its event horizon is the exact point of no return.

The Sun's mass is too small for it to collapse into a black hole of its own accord. But other stars can, and Cygnus X-1 was one of them. The only reason that we can see anything at all in its place is that the black hole isn't alone but being orbited by a huge star at close proximity. This hot, restless supergiant is continuously flinging material out of its atmosphere and in the direction of the black hole, where it disappears beyond the event horizon. Though not without leaving a trace: As it races toward the black hole at a dizzying speed, along an ever narrowing spiral, it becomes intensely hot and starts to shine—shining with particular intensity in the invisible X-ray and ultraviolet regions of the spectrum.

Using suitable telescopes, we were able to confirm this radiation in the region of Cygnus X-1 and watch the matter from the supergiant stop glowing and disappear the instant it crossed the event horizon. Cygnus X-1 was the first black hole whose existence we were able to prove beyond the shadow of a doubt. Incidentally, it poses no danger to us at all: Unless you get too close to their event horizons, black holes are relatively harmless. Given that it's 6,100 light-years away, Cygnus X-1 is a fascinating research subject we can calmly contemplate from a safe distance.

91

THE GREEN STAR
It's Black and White

THERE'S NO SUCH THING as a green star. We can see any number of yellow, red and blue stars in the sky—assuming, that is, that the conditions are favorable and our eyes are good enough; otherwise we only perceive them as white. Yet a green star has eluded us so far, because a green star has to be white, since it's a black body.

That may sound ludicrous but it's perfectly proper astronomy. Black bodies are objects that completely absorb any radiation they encounter, and whose temperature alone (and not, say, the constitution of their surface, or any other aspects of their makeup) determines the way in which they themselves radiate energy. Black bodies are a concept from theoretical physics and have no actual counterpart in nature, but it's still possible to think of real stars as just such black bodies.

What this means in practice is that a star always radiates energy in all colors of the rainbow. For example, when you see a red star, that star is also emitting blue, yellow and green light. The same goes for blue stars. What determines whether a star appears to us as blue, red or yellow is where in the spectrum its peak radiation falls. Blue stars are hotter than red ones, so they emit most of their energy at blue wavelengths, while cooler stars emit most of their light within the red region of the spectrum.

The peak radiation of stars like the Sun, which are cooler than blue stars but warmer than red ones, falls within the green region of the spectrum. The reason we don't perceive the Sun as green but yellowish white is that although the receptors in our eyes can discern blue, green and red light, their sensitivities overlap slightly. The Sun sends us light in every hue—a lot of green, but also red, yellow and blue—but when it meets our eyes the receptors are unable to clearly distinguish between the various proportions. We consequently perceive a blend of colors, which in the Sun's case is white.

The same happens with stars whose radiation peaks in the red or blue region. In their case, however, we have to remember that there are certain colors that our eyes can't perceive at all—infrared, ultraviolet and all those other colors in the electromagnetic spectrum that we can "see" with measuring instruments, but for which our eyes have no receptors. A blue star's radiation peaks a lot closer toward the ultraviolet end of the spectrum than that of a "green" star like the Sun, which means that a blue star emits more ultraviolet light than the Sun. Yet light at those wavelengths is invisible to the receptors in our eyes, so all we see is the intense blue portion of it, which combines with the less intense green, yellow and red portions to produce a comparatively blueish light. In the same way, stars whose radiation peaks in the red region emit far more infrared radiation than blue or "green" stars, but our eyes only register the intense red, mixed in with the fainter green, yellow and blue portions.

To our eyes, then, starlight appears only as a blend of every hue it radiates, and because green light falls exactly in the middle of the visible spectrum, it's always transmuted by the other colors. For a star to actually look green, it would have to radiate energy only in the form of green light, and no other color whatsoever. And that's physically impossible.

92

GLIESE 710

A Close Encounter in the Distant Future

ONE THING THERE'S MORE THAN ENOUGH OF in space is nothingness. Only here and there will you find a little something in the nothing. For instance a galaxy, but then again, galaxies are also chiefly made of nothing—and of stars, separated by unimaginable distances.

There are 4.2 light-years between the Sun and its next closest star, Proxima Centauri; that's just under an almost inconceivable 25 trillion miles (40 trillion kilometers). And then there's the star Gliese 710, which is just 62 light-years from Earth and doesn't rank even among the 1,000 nearest to us.

But one day this will change. Right now, we still have to make quite an effort to spot Gliese 710 up there in the sky, in the Serpens constellation. The star is too feeble to be apprehended without the aid of technology; it's two thirds the size of the Sun, and its mass, too, is little more than half the Sun's.

Gliese 710 would hardly command our attention were it not for one thing, which astronomers discovered back in the twentieth century. Gliese 710 is traveling toward us. There's no cause for panic just yet, because it isn't due to arrive for another 1.35 million years, and even when it does Gliese 710 will collide neither with the Sun nor with us. Although our latest measurements suggest that it'll be a close call, astronomically speaking.

We can't predict the motion of a star over such long periods of time with absolute accuracy, but the way things look at the moment, Gliese 710 is going to come within 15,000 astronomical units of Earth. Which is about 0.2 light-years, or 1 trillion miles (about 2 trillion kilometers)—enough of a buffer, considering that the distance between Earth and the Sun amounts to one astronomical unit, and Neptune, the most remote planet in the solar system, is only thirty astronomical units away.

So Gliese 710's direct influence on our planet will be limited. Indirectly, however, this close encounter in the distant future could well have dramatic consequences. Our solar system is surrounded by a gigantic cloud of asteroids and comets, the Oort cloud, which at its farthest extends to about 100,000 astronomical units; and Gliese 710 will be flying right through it. Normally, the cloud's trillions of boulders don't affect us all the way over here, inside the solar system. They're far away from Earth, and likely to stay there, unless an external disturbance forces them to depart their harmless orbits and set off in the direction of the Sun.

Gliese 710's gravitational influence would constitute just such an external disturbance and could result in an increase in the number of comet and asteroid impacts on the planets of our solar system. Of course, it's too soon to predict how many of them would hit Earth. But if humans are still around in that distant future, they'll hopefully have long ago developed an asteroid defense system. On the plus side, they'll be able to see Gliese 710 in the sky without a telescope. At night, it'll outshine Mars—and remind us never to take our eyes off the stars, even if the universe is dominated by nothingness.

93

GRB 080319B

The Biggest Explosions in the Universe

ON MARCH 19, 2008, we were able to see farther into the universe than ever before with the naked eye. At least, we could for thirty seconds—and only if we happened to be looking in the right place at the right time. On that day a weak new star appeared in the Boötes constellation, which you could just about make out without optical aid. Yet this "new star" was in fact the most immense explosion ever recorded. Its light had traveled across the universe for 7.5 billion years. The event that occurred in this remote galaxy all those years ago is known as a "gamma-ray burst," and it is somewhat disconcerting.

In 1967, those whose job it was to evaluate the observational data received from the Vela satellites also got a fright. The US was using Vela to monitor compliance with the Nuclear Test-Ban Treaty signed by the US, USSR, and UK in 1963. The radioactivity of Earth's atmosphere had increased to a worrying degree in previous years, and no one was supposed to test atomic bombs anymore, at least not aboveground. The Vela satellites had been deployed into space to ensure that everyone was keeping their word; an aboveground atomic explosion would release a massive amount of high-energy radiation, which would be picked up by the satellites' measuring instruments.

On July 2, 1967, they observed just such a blast of so-called gamma radiation—and a colossal one. However, once the first shock had passed, it quickly became clear that nobody here on Earth had flouted the terms of the Nuclear Test-Ban Treaty. The explosion was more powerful than anything an atomic weapon could have produced, and it hadn't happened on Earth, but somewhere in space. Over the following years, Vela observed more and more such gamma-ray burst events. Nobody was sure what was causing them, and the general confusion only increased when astronomers realized that the explosions were happening deep in space, and that their light was coming to us from galaxies a very long way from here.

All they knew for sure was that whatever was exploding there, it was radiating more energy in a handful of seconds than our Sun has during its entire lifetime of several billion years. Astronomers kept seeing more and more gamma-ray bursts and were soon able to distinguish two types: those that would flare up for just a few seconds, and those that would blaze for half a minute or more. We now know that they are actually two different phenomena.

The longer bursts are most likely produced by "hypernovas." Like supernovas, hypernovas are large stellar explosions, but in their case the star is tremendously big, and undergoing a correspondingly tremendous explosion. The inside of an extremely massive dying star can very quickly collapse under its own weight straight into a black hole. When this happens, the star's outer layers start spinning very rapidly around the freshly created hole and become so hot that they start producing powerful gamma radiation, which they emit into space in the form of high-intensity rays.

The shorter bursts, on the other hand, originate in collisions between two neutron stars; neutron stars form when slightly less massive stars die, and they're extremely compact, containing a mass equivalent to the Sun's inside a space measuring just a couple of miles across. When two of them collide, they transform into a black hole, which then also releases gamma radiation.

The gamma-ray bursts we've observed to date have all occurred in remote galaxies and are no threat to us. An explosion like that can only have consequences for Earth if it happens less than about 3,000 light-years away. Its gamma-ray burst could then destroy Earth's ozone layer, which protects us from the Sun's destructive UV radiation. Perhaps an event like that was responsible for the so-called Ordovician mass extinction 443 million years ago, when more than 85 percent of Earth's species vanished. So far there's no evidence to suggest that a gamma-ray burst has actually happened in our cosmic neighborhood. And we know of no sufficiently large star in our vicinity that could produce a hypernova in the near future, and thereby endanger life on Earth.

94

GW150914

Gravity's Light

"GW150914" REFERS NOT TO A STAR but to an event that, on September 14, 2015, inaugurated an entirely new kind of astronomy. On that day, we finally managed to achieve something we'd been trying to do for decades: the first direct measurement of gravitational waves.

The phenomenon's existence was predicted by Albert Einstein in 1916. One unexpected consequence of the general theory of relativity he'd published shortly before was an entirely novel way of looking at the structure of space. Space wasn't just there; it could be distorted. Every mass bends space-time, and we experience these distortions as the gravitational force of attraction. Furthermore, according to Einstein, any changes in the curvature of space can only unfurl at the speed of light. The gravitational waves whose existence he posited are the ripples caused by these changes in curvature traveling through space.

What causes them is a system of rapidly moving masses, for example two celestial bodies orbiting each other. As Earth revolves around the Sun, it continually emits gravitational waves, like ripples around a pebble cast into a pond. But the waves produced by Earth are too faint for us to detect.

They never are an easy thing to detect. When a gravitational wave meets Earth, all space around it changes—in simple terms, Earth is first squeezed a little, then stretched a little. But because the whole of space changes, so does everything else in it: All lengths, all distances, everything is affected by gravitational waves, including us and our measuring instruments.

To track them down, you therefore have to build a very particular kind of detector. For this you need a laser that can send light simultaneously in two different directions; when the beams have reached a specific distance, they are reflected and returned to their point of origin. If their journeys were equally long, they should get back at exactly the same time. However, if at that very moment a gravitational wave happens to run through the detector, space expands in one direction but contracts in the other, and because light always moves at a constant speed the two beams are no longer synchronized.

The principle is straightforward but constructing a functioning gravitational wave detector took decades. The distance covered by the light has to be measured to an accuracy smaller than the diameter of an atomic nucleus. When we finally accomplished the feat, our search bore fruit surprisingly quickly: Even as the team was completing the first test runs in September 2015, they found the signal they'd been looking for. It originated from a collision between two black holes, one thirty-six times and the other twenty-nine times the Sun's mass. They were orbiting each other at close quarters, and thereby producing gravitational waves powerful enough to be registered by the detector.

During the last 0.2 seconds before the two black holes collided, the speed at which they were spinning around each other increased from 30 to 60 percent of the speed of light. In these final moments of their existence, the celestial bodies caused space-time to oscillate so much that we were able to measure it even at the distance of 1.4 billion light-years that separates us from the collision site. Since this first-ever observation of gravitational waves (which was awarded the 2017 Nobel Prize in Physics), we've detected nearly a dozen such events, and many more are sure to follow.

Gravitational waves have provided us with a whole new way of conducting research. Before then, we had to rely on light—i.e., on electromagnetic radiation—whenever we wanted to collect information about space. Now we can also observe the universe using gravitation, and "see" events that are invisible in normal light. We can investigate the behavior of black holes, and future gravitational wave detectors may in addition help us to analyze events that occurred immediately after the Big Bang: The particular way in which it made the universe oscillate could hold the key to how it was formed. "Gravity's light" signals the dawn of a new era in astronomy.

95

R136a1

The Monster in the Tarantula Nebula

FOR A RECORD HOLDER, the star R136a1 has a rather dull name. Fortunately, though, it belongs to the rather more strikingly named Tarantula Nebula, which means that we can—as I have done in the subtitle of this chapter—address it with the respect it deserves. R136a1 is, of all the stars known to astronomy, the brightest and most massive. It shines ten times brighter than the Sun, and the only reason we don't notice it is that it's so very remote. Together with the cosmic clouds of the Tarantula Nebula that surrounds it, it's located nearly 180,000 light-years from us in the Large Magellanic Cloud, one of our neighboring galaxies.

If R136a1 was where the Sun is, it would exceed the Sun's brightness by as much as the Sun's exceeds the Moon's. R136a1 is a whole 265 times heavier than the Sun, and if it really was the center of the solar system, the massive increase in gravitational force would shorten Earth's orbit from 365 days to a mere 21.

When this mighty star was formed, it was even heavier than that. Unlike us humans, stars lose weight with age; the greater a star's mass, the greater the pressure the mass exerts on its core, and the higher the prevailing temperatures; and the hotter the star is, the more quickly the atoms of the gas of which it consists move—both inside the star and on its surface. Whereas the Sun has a surface temperature of only about 10,000 degrees Fahrenheit (5,000 degrees Celsius), R136a1's is a scorching 72,000 (40,000); the speedy atoms can no longer be held back by the star's gravitational force and stream out into space. This "stellar wind" causes the star to lose more and more mass the hotter it grows. R136a1 has already shed 35 solar masses; at birth, it weighed as much as 300 times more than the Sun.

By astronomical standards, though, this colossus isn't all that old yet. It was formed a mere million years ago—a veritable cosmic baby. Yet it'll never reach the grand old age of the Sun, which has been sending its light into space for 4.5 billion years, because the greater a star's mass, the hotter it is, and the faster it burns through its material. The motto "live fast, die young" goes for actual stars as much as 1950s Hollywood actors. R136a1 is unlikely to survive much longer than another few dozen million years.

The fact that it exists at all is astonishing enough. The hotter it is, the more radiation a star emits; the radiation starts pushing against the stellar mass surrounding it, and eventually becomes too powerful for the gravitational force, which has so far managed to keep it in check. The star is torn apart by its own radiation, so to speak, even before it has really started living. Until the discovery of R136a1 in 2010, astronomers had assumed that no star can have more than 150 times the mass of the Sun. But the monster in the Tarantula Nebula has forced them to change their minds: There are evidently exceptions to the rule. Or perhaps there are other ways for stars of that size to be produced—for example by a merger of several young stars born close together in the same cluster.

We obviously still have some research left to do before we can hope to fully understand the very brightest of stars. In the meantime, though, perhaps we can come up with a better name for R136a1.

96

TRAPPIST-1

The Perpetual Discovery of the Second Earth

IN 2017, SOME SECTIONS of the media reported that a "second Earth" had been discovered at last. The planet they were referring to orbits the star with the designation Trappist-1, and the illustrations that accompanied the reports depicted an ocean with ice floating on it, stretched out under a sky resplendent with a red sun and other planets. Yet the "second Earth" thus portrayed exists only in the artists' imagination; not only that, but Trappist-1's planet is far from the first "second Earth" ever discovered. We'd already discovered one of these orbiting the star HD 85512 in 2011, and another in 2012, this one orbiting GJ 667C. And another in 2014, orbiting Kepler-186, and yet another in 2015, orbiting Kepler-452. In 2016, a "second Earth" was found orbiting Proxima Centauri. In short, the media has regularly announced their discovery for years.

Our desire to find such a planet is understandable. Ever since antiquity, humans have wondered whether there are other worlds like Earth in the universe, but it wasn't until 1995 that we finally discovered a planet orbiting a star other than the Sun. However, in the more than 2,000 years before that and the decades since, one question has always been at the center of our search for alien worlds: Is there a planet out there on which we humans can live? Is Earth a solitary oasis in a life-averse cosmos, or is there somewhere else with the right conditions to sustain life?

None of the planets discovered by astronomers in the last few decades are a "second Earth." They are too large, too hot, too cold, or otherwise inimical to habitation. During the first few years, the available technology was in any case only capable of detecting very large planets—gas giants like Jupiter and Saturn, which don't have a solid surface and can therefore definitely not be a "second Earth." Only when we discovered the planet CoRoT-7b in 2009 did we know for sure that we were dealing with a celestial body whose surface is solid. It was the first known "terrestrial planet," and the fact that astronomy chose to designate it as such is perhaps partly responsible for this media glut of "second Earths."

A "terrestrial" or "earthlike" planet is one that has a solid surface and consists chiefly of layers of rock enveloping a metal core. In our solar system, this is true not only of Earth, but also of Mercury, Venus and Mars, though we probably wouldn't normally call them "earthlike." All three are extremely inimical to life—though they do resemble Earth as far as their interior constitution is concerned.

Even if a planet is "earthlike" in astronomical terms, there's no reason to suppose that earthlike life is to be found on it. No more than a handful of the terrestrial bodies we know of so far is even located in the "habitable zone" of their stars—i.e., where it isn't too hot or too cold, and where life-producing conditions could thus exist on their surface. "Could" is the operative word, because we don't know whether they actually exist anywhere. We know too little about them to be able to tell. Until we are in a position to analyze the atmospheres of other planets—or even to find out whether they have one in the first place—and until we can examine their magnetic fields and geological processes, we won't know anything about the conditions on the ground. Only the next generation of space telescopes and so-called large telescopes will make this possible. These instruments, such as the Extremely Large Telescope being built by the European Southern Observatory in Chile—due to be completed in the mid-2020s—will be able to deliver much more detailed data about the planets of other stars.

Till then, the "second Earth" will remain an item of mis-, or misunderstood, information. But we've already waited for 2,000 years, and now we won't have to wait much longer; if there is such a thing as a real second Earth out there, we'll find it in the next few decades. Trappist-1 isn't a bad place to start looking: This small star, just forty light-years away from us, is orbited by no fewer than seven earthlike planets. Three of them are inside the habitable zone. Perhaps the fanciful illustrations in those media reports will belatedly turn out not to have been entirely wrong after all.

P CYGNI

A Question of Distance

HOW FAR CAN WE SEE? Assuming that nothing obscures our view and the atmosphere is unpolluted, the maximum distance is 3 miles (5 kilometers), at which point the eyes of a person of average height will meet the curvature of Earth's surface. But what if you look straight up? Which of the stars we can see in the sky above us on a clear night is farthest away?

It's an obvious question to ask, but answering it is far from easy. The word "see" itself already proves a stumbling block: The better the optical instruments at your disposal, the farther you can see; if your telescope is good enough, you can even make out unbelievably remote galaxies—but that's not really what we normally mean by "see." What we mean, rather, is "see with our own eyes," and everyone's eyes are different. Not only that, but the conditions under which we deploy them also play a role. Right in the middle of a large, brightly lit city at night, we'll naturally make out fewer stars than on a dark night somewhere in a lonely desert.

There's also the problem of determining distance: To see stars is easy; to measure their precise distance is not. We don't yet know the exact distance of all the stars we can see with the naked eye—far from it—so we can't even offer an unequivocal answer to which one of *them* is the most remote.

One candidate could be the star HD 70761, which is just about bright enough to be visible without the aid of technology, so long as the conditions are perfect and your eyes excellent. But aside from the fact that it exists and may be a lot more than 10,000 light-years away, nothing much is known about this celestial body. V509 Cassiopeiae is a bit more interesting—it's a yellow giant star (possibly) more than 8,000 light-years away. It's nearly 1,000 times bigger and 350,000 times brighter than the Sun, and much easier to see than HD 70761.

But if there is no straightforward answer to be had yet, we might as well crown P Cygni the winner in the interim. At a distance of about 7,000 light-years it's reasonably far away, and there's much to say about it. Although we can nowadays see it fairly easily with the naked eye, no one noticed it until August 18, 1600, when it was discovered by the Dutch cartographer Willem Blaeu. The star appeared suddenly and very brightly in the Cygnus constellation, but quickly darkened again, and by 1626 you couldn't see it at all except through a telescope. During the following decades it sometimes shone brighter, sometimes less brightly, and only from 1715 onward did its brightness slowly increase to its current intensity.

P Cygni is a "luminous blue variable" (LBV). This is what we call hot blue stars whose life is nearly over. They have the greatest mass a star can possibly have, which is why they're so enormously hot and luminous. P Cygni is more than half a million times brighter than the Sun, and its radiation is so powerful that it's literally being torn apart by it. The star keeps casting off bits of its atmosphere and outer gas layers, and this is what's causing the fluctuations in its brightness. An LBV can only withstand this chaotic phase for a few tens of thousands of years before dying in a powerful explosion.

Considering that P Cygni will cease to exist in the near future—"near" in astronomical terms, that is—there's no reason why we can't in the meantime award it the title of "Most Distant Star Visible Without Optical Aid."

98

OUTCAST

Through the Milky Way at Hyperspeed

OUTCAST IS RACING THROUGH SPACE at 441 miles (709 kilometers) a second. Which is quick, even for a star. So quick, in fact, that the gravitational force of attraction of the entire Milky Way, with its billions of stars, isn't capable of holding it back. In the distant future Outcast, which has already reached the edge of our galaxy, will leave us far behind and enter the emptiness of intergalactic space.

Outcast is a hot young star. It's only 350 million years old and has experienced a fair amount in its—for a star—relatively short life. It was probably born with a partner, as part of a binary system, somewhere close to the galactic center. Things are a tad more hectic there than in the suburbs of the Milky Way that we and the Sun call home. At the center of the Milky Way the stars are a lot closer together, and if you're not careful you'll come across the supermassive black hole.

This mighty object sits in our home galaxy, and there are others inhabiting all large galaxies. It has 4 million times the Sun's mass and is influencing the entire Milky Way's development with its gravitational force, and with the high-energy process occurring in its vicinity. About 100 million years ago, it also sealed Outcast's fate.

We know that, back then, Outcast must have been loitering near the Milky Way's center. We know this because we can tell that it's moving in the opposite direction from the center, as if it were desperate to get away from it as fast as possible. And that is, in fact, what it's doing.

Once upon a time, it and its fellow star orbited each other, and together revolved around the central supermassive black hole in the middle of the Milky Way. But once you get close to a black hole, there's no getting away from it—which is exactly what happened to Outcast's fellow, while Outcast itself only escaped by the skin of its teeth. The binary system was ripped apart in close proximity to the black hole by the hole's enormous force of attraction. The other star was swallowed by the hole; Outcast was flung out like a hammer in the hammer throw.

At least that's what astronomers thought had happened when they first spotted Outcast in 2005. It was the first, and at the time only, known "hypervelocity" star, which is what we call stars that are fast enough to overcome the Milky Way's gravitational pull. In the interval, however, we've discovered almost two dozen of them, and not all have the black hole to thank for their speed. Some of them are coming from the wrong direction or are much too young; even allowing for their high velocity, they wouldn't have been able to cover the distance between the central black hole and their current position within their lifetimes. So there have to be other mechanisms that can cause stars to speed up to such enormous velocities.

But we can't quite agree on what these are. It might be powerful supernovas, or perhaps an encounter between the Milky Way and another galaxy—such encounters used to happen frequently, and if that's the case, and if the two galaxies got very close to each other, the Milky Way could have relieved the other one of a couple of stars by means of its gravitational force. However, there may be an entirely different explanation, which we are yet to discover.

99

S MONOCEROTIS

The Spiral Galaxy in the Christmas Tree

IT'S ACTUALLY A LITTLE EMBARRASSING to think how long it's taken us to discover the shape of our own Milky Way. We spent decades admiring remote galaxies in all their glory, and yet hadn't the faintest idea what our own cosmic home looks like. You might say that we couldn't see the forest for the trees. From here, among all those billions of stars in the Milky Way, it was simply impossible to obtain a proper overview of the thing. Until Boxing Day in 1951, that is, when the American astronomer William Morgan caused a lecture hall filled with usually rather reticent scientists to break out in wild applause and much enthusiastic stomping of feet. He had just presented them with the first-ever images showing the structure of the Milky Way. They included the star S Monocerotis, fittingly located in a region of the sky called the Christmas Tree Cluster.

For a long time, it was thought that the Milky Way was nothing but a flat assembly of stars, a disk without any notable structure. But then, in the 1920s, the astronomer Edwin Hubble realized that those spiral nebulas in the sky were actually remote spiral galaxies. They weren't part of the Milky Way, but gigantic agglomerations of stars separated from us by equally gigantic tracts of empty space. We naturally wanted to know whether we ourselves were living in such a galaxy, too, one whose stars are arranged in a spiral formation. All we had to do was measure the distances between us and the stars and mark them on a three-dimensional chart.

But that's easier said than done. For one thing, there are far too many stars—to this day, we've only measured the exact distances of about a billion out of the 200 billion stars in the Milky Way. For another, measuring their distance doesn't only become more difficult the farther away a star is, but some remote stars are practically invisible; they were especially hard to see in the first half of the twentieth century, when telescopes weren't what they are today.

Nevertheless, William Morgan was keen to know what the Milky Way looks like and came up with a pragmatic solution: He would measure only the distances of the really bright stars, those he could observe even at a great distance. From previous studies of other spiral galaxies, he knew that bright, young, blue-white shining stars such as these usually form in the spiral arms of galaxies. S Monocerotis is one such star—or rather, it's stars, plural: a binary system consisting of two celestial bodies, both of which have a mass greater than twenty times that of the Sun.

There was one more problem to overcome: The cosmic dust that exists everywhere between the stars of the Milky Way blocks out part of the light, and thus makes it difficult to determine how far away the stars are. It isn't an insurmountable obstacle, though, because the blue portion of starlight is more significantly affected by this dust than the red portion. By comparing a star's brightness in these two colors you can work out how much the dust is influencing its light.

This only works if you know in advance exactly what color a given star is. Quite bluish? Not so bluish? Bluish white? Just white? Yet this is the very information concealed by the dust. However, there's another way to discover the color of a star—for it to work, you have to know how hot the star is, which you can tell from its chemical composition. Astronomers had been classifying stars in this way since the beginning of the twentieth century, but hadn't yet fully grasped the "spectral types" relevant to the bright young things Morgan was most interested in. So he first had to perform the laborious task of working out how to assign these stars their rightful places in the existing system.

Which he did; and in 1951 he assembled the observational data of S Monocerotis and other stars into an image that showed two spiral arms. We, too, then, live in a spiral galaxy. But what it looks like in detail, and how many spiral arms it has in total, remains unclear—the trees are still getting in the way of the forest.

100

ZETA OPHIUCHI

Cosmic Rays and Climate Change

SPACE IS FILLED with cosmic radiation. This radiation consists of particles such as electrons or protons, which come from everywhere: Some originate in the Sun, others in other stars, some have even traveled from distant galaxies. Cosmic radiation is released by supernovas, but is also sent out into space by stars during the normal course of their lives, together with their light.

We on Earth receive most of our cosmic radiation from the Sun, but the Sun also shields us from the cosmic radiation reaching us from the outside. It pushes outward, so to speak, with its own particle radiation, the so-called solar wind, creating a barrier that blocks out "alien" cosmic radiation. How well this works depends on how strong the solar wind is at any given time. When cosmic rays hit Earth, the particles inside them act on the aerosols in the air around it; aerosols are small particles—rust, pollen, dust, bacteria and a number of other things—that drift about and can become electrically charged when they interact with cosmic radiation. This makes it easier for the water droplets in the air to accumulate around them and form large clouds; and the more cosmic radiation there is, the more cloud is produced. Therefore, any change in cosmic radiation inevitably causes a change in the amount of cloud in the atmosphere, and consequently also changes Earth's climate. So climate change has nothing to do with the greenhouse gases we humans are expelling into the atmosphere; rather, it is being caused by entirely natural variations in solar wind and cosmic radiation.

This is the story often told by those who refuse to accept the findings of climate research, and who can't be bothered with climate action. They can call on a real physicist for support, too: Henrik Svensmark, the author of a 1997 theory concerning the influence of cosmic radiation on climate. He expands on his theory in a book from 2007 (entitled *The Chilling Stars*), and among other things cites the star Zeta Ophiuchi as proof for his claims.

Zeta Ophiuchi is one of the hundred brightest stars in the night sky, currently located 360 light-years away from us and moving extremely fast. It appears that something gave it an almighty shove about a million years ago. At the time, it was still part of a binary system, but its fellow exploded in a supernova and sent Zeta Ophiuchi off on its high-speed flight through the Milky Way. According to Svensmark, a million years ago this supernova, which occurred comparatively close to the solar system and produced a vast amount of cosmic radiation, resulted in an ice age on Earth; he also contends that all other ice ages were triggered by exploding stars. Therefore, he argues, the blame for the current climate change lies not with us humans, but with the solar wind's currently well-oiled protective operations.

It's no surprise that climate change deniers have pounced on the argument. It's true that it isn't entirely implausible physics—the mechanism is theoretically possible. However, over the years numerous researchers have again and again checked and re-checked Svensmark's theory and concluded that cosmic radiation plays a far smaller role in the formation of clouds than Svensmark assumed. CERN—the European Organization for Nuclear Research—has even simulated the formation of clouds in laboratory experiments, using cosmic radiation produced by particle accelerators; they also found that, although the mechanism exists, it's much too weak to have any significant impact on climate change.

If you add this to the discoveries made by climate scientists during the last few decades, it's abundantly clear that we humans are causing Earth to grow warmer. We have to protect our climate ourselves—we, not the stars, are responsible for our actions.

More Stories About the Universe

MY HISTORY OF THE UNIVERSE has come to an end. It's the version I wanted to tell, but of course there are countless more facts and myths about the skies, more protagonists and more stories, which haven't found their way into this book. Everyone has their own way of looking at the stars, and I hope that this book will inspire as many readers as possible to take a look at the stars and contemplate the universe.

If you need a little help observing the nocturnal sky, you can find it for example in *The Stars: A New Way to See Them* and *Find the Constellations* by H. A. Rey; they were actually written for children, but are just as enjoyable for adults. A slightly more sober, but highly informative, way to learn about what happens in the sky can be had with the annual *Guide to the Night Sky* published in association with the Royal Observatory Greenwich. Emily Winterburn's *The Stargazer's Guide: How to Read Our Night Sky* not only does what the title promises, but also contains many of the wonderful myths we tell each other about the stars in the sky; you'll also find these in Susanna Hislop and Hannah Waldron's *Stories in the Stars: An Atlas of Constellations*, *Star Stories* by Anthony Aveni and *Seeing Stars: A Complete Guide to the 88 Constellations* by Sara Gillingham. If you'd like to immerse yourself in the history of the constellations, it's worth taking a look at John C. Barentine's slightly more academic *The Lost Constellations: A History of Obsolete, Extinct or Forgotten Star Lore*.

I have tried to include as many subject areas of astronomy as possible in this book, to show just how multifaceted and vibrant our preoccupation with the sky has always been. Nevertheless, this book provides but a brief glimpse into a universe filled with too many stories for one person alone to tell. Fortunately, there are many others who have written their own stories about the universe, and I urge any interested readers to take a look at the little list of useful titles I have appended to this book. It may not be comprehensive, but it does include many of the books I myself referred to while writing this one, and others that I think are well worth reading. But don't let me keep you from setting off on your own voyage of discovery into the universe of books—there's no such thing as too many stories about the universe!

Acknowledgments

My deepest gratitude to my former editor Christian Koth and my new editor Annika Domainko. Without their help, this book would by no means be as good as I hope it is now. Thank you also to Dagmar Fuchs, Ruth Grützbauch, Matthias Kittel, Dieter Kreuer and Elke Pehamberger-Müllner, who read this book at various stages of its inception and whose insightful comments guided my improvements here and there. The greatest thank-you, of course, is reserved for everyone who looked up at the sky over the past thousands of years and discovered all the things we know about the universe today. Without them, these stories wouldn't exist. And I hope that humans will never tire of telling each other stories about the stars.

Further Reading

General and Popular Science Books about Astronomy

Couper, Heather, and Nigel Henbest, *The Secret Life of Space* (London: Aurum Press, 2015).

Freistetter, Florian, *Der Komet im Cocktailglass: Wie Astronomie unseren Alltag bestimmt* ("The Comet in a Cocktail Glass: How Astronomy Affects Our Daily Lives") (Munich: Hanser Verlag, 2013).

Hawking, Stephen, and Leonard Mlodinow, *The Grand Design: New Answers to the Ultimate Questions of Life* (London: Bantam Press, 2010).

Oberhummer, Heinz, *Kann das alles Zufall sein? Geheimnisvolles Universum* ("Can it Really Be Just a Coincidence? The Mysterious Universe") (Salzburg: Ecowin Verlag, 2008).

Puntigam, Martin, Heinz Oberhummer and Werner Gruber, *Das Universum ist eine Scheißgegend* ("The Universe Is a Dump") (Munich: Hanser Verlag, 2015).

Singh, Simon, *Big Bang: The Most Important Scientific Discovery of All Time and Why You Need to Know About It* (London: Fourth Estate, 2004).

Biographies

Freistetter, Florian, *Isaac Newton: The Asshole Who Reinvented the Universe*, trans. Brian Taylor (New York: Prometheus Books, 2018).

Haramundanis, Katherine, ed., *Cecilia Payne-Gaposchkin: An Autobiography and Other Recollections*, 2nd ed. (Cambridge: Cambridge University Press, 1996).

Johnson, George, *Miss Leavitt's Stars: The Untold Story of the Woman Who Discovered How to Measure the Universe* (New York: W. W. Norton, 2005).

Kerner, Charlotte, and Doro Göbel, *Sternenflug und Sonnenfeuer: Drei Astronominnen und ihre Lebensgeschichte* ("Starflight and Sunfire: Three Astronomers and Their Life Stories") (Weinheim/Basel: Beltz & Gelberg, 2004).

Rublack, Ulinka, *The Astronomer and the Witch: Johannes Kepler's Fight for His Mother* (Oxford: Oxford University Press, 2015).

Samhaber, Friedrich, *Der Kaiser und sein Astronom: Friedrich III und Georg Aunpekh von Peuerbach* ("The Emperor and His Astronomer: Frederick III and Georg von Peuerbach") (Peuerbach, Austria: Municipality of Peuerbach, 1999).

Sobel, Dava, *A More Perfect Heaven: How Copernicus Revolutionized the Cosmos* (London: Bloomsbury, 2011).

Winterburn, Emily, *The Quiet Revolution of Caroline Herschel: The Lost Heroine of Astronomy* (Stroud, UK: History Press, 2017).

History of Astronomy

Al-Khalili, Jim, *Pathfinders: The Golden Age of Arabic Science* (London: Allen Lane, 2010).

Andersen, Johannes, David Baneke and Claus Madsen, *The International Astronomical Union: Uniting the Community for 100 Years* (Berlin: Springer, 2019).

Bartusiak, Marcia, *The Day We Found the Universe* (New York: Pantheon, 2009).

Clark, Stuart, *The Sun Kings: The Unexpected Tragedy of Richard Carrington and the Tale of How Modern Astronomy Began* (Princeton, NJ: Princeton University Press, 2007).

Drößler, Rudolf, *2000 Jahre Weltuntergang: Himmelserscheinungen und Weltbilder in apokalyptischer Deutung* ("2,000 Years of the End of the World: Apocalyptic Interpretations of Celestial Apparitions and Images of the World") (Würzburg, Germany: Echter, 1999).

Freely, John, *Before Galileo: The Birth of Modern Science in Medieval Europe* (London: Duckworth, 2013).

Hirshfeld, Alan, *Parallax: The Race to Measure the Cosmos* (Mineola, NY: Dover Books, 2001).

Meller, Harald, and Kai Michel, *Die Himmelsscheibe von Nebra: Der Schlüssel zu einer untergegangenen Kultur im Herzen Europas* ("The Nebra Sky Disc: The Key to a Vanished Culture at the Heart of Europe") (Berlin: Ullstein, 2018).

Oeser, Erhard, *Die Suche nach der zweiten Erde: Illusion und Wirklichkeit der Weltraumforschung* ("The Search for the Second Earth: Illusion and Reality in Space Exploration") (Darmstadt, Germany: Wissenschaftliche Buchgesellschaft, 2009).

Wulf, Andrea, *Chasing Venus: The Race to Measure the Heavens* (London: William Heinemann, 2012).

Planets and Planetary Systems

Brown, Mike, *How I Killed Pluto and Why It Had It Coming* (New York: Spiegel & Grau, 2010).

Freistetter, Florian, *Die Neuentdeckung des Himmels* ("The Rediscovery of the Skies") (Munich: Hanser Verlag, 2014).

Sasselov, Dimitar, *The Life of Super-Earths: How the Hunt for Alien Worlds and Artificial Cells Will Revolutionize Life on Our Planet* (New York: Basic Books, 2012).

Schneider, Reto, *Planetenjäger: Die aufregende Entdeckung fremder Welten* ("Planet Hunters: The Exciting Discovery of Alien Worlds") (Basel, Switzerland: Birkhäuser, 1997).

Astronomical Anecdotes

Ashbrook, Joseph, *The Astronomical Scrapbook: Skywatchers, Pioneers and Seekers in Astronomy* (Cambridge: Cambridge University Press, 1985).

Baum, Richard, *The Haunted Observatory: Curiosities from the Astronomer's Cabinet* (New York: Prometheus Books, 2007).

Hirshfeld, Alan, *Starlight Detectives: How Astronomers, Inventors, and Eccentrics Discovered the Modern Universe* (New York: Bellevue Literary Press, 2014).

Tucker, S. D., *Space Oddities: Our Strange Attempts to Explain the Universe* (Stroud, UK: Amberley, 2017).

Cosmology, Black Holes and Gravitational Waves

Baggott, Jim, *Quantum Space: Loop Quantum Gravity and the Search for the Structure of Space, Time, and the Universe* (Oxford: Oxford University Press, 2018).

Bartusiak, Marcia, *Black Hole: How an Idea Abandoned by Newtonians, Hated by Einstein, and Gambled on by Hawking Became Loved* (New Haven, CT: Yale University Press, 2015).

Collins, Harry, *Gravity's Kiss: The Detection of Gravitational Waves* (Cambridge, MA: MIT Press, 2017).

Freistetter, Florian, *Stephen Hawking: His Science in a Nutshell*, trans. Brian Taylor (New York: Prometheus Books, 2020).

Greene, Brian, *The Fabric of the Cosmos: Space, Time, and the Texture of Reality* (New York: Alfred A. Knopf, 2003).

Hawking, Stephen, *A Brief History of Time: From the Big Bang to Black Holes* (London: Bantam Press, 1988).

Hawking, Stephen, *The Universe in a Nutshell* (London: Bantam Press, 2001).

Hossenfelder, Sabine, *Das hässliche Universum* ("The Ugly Universe") (Frankfurt am Main: S. Fischer Verlag, 2018).

Levin, Janna, *Black Hole Blues and Other Songs from Outer Space: Black Holes and the Quest to Hear the Invisible* (London: Bodley Head, 2016).

Rovelli, Carlo, *Seven Brief Lessons on Physics* (London: Allen Lane, 2015).

Smoot, George, and Keay Davidson, *Wrinkles in Time: Witness to the Birth of the Universe* (New York: William Morrow, 1994).

Vaas, Rüdiger, *Signale der Schwerkraft: Gravitationswellen: Von Einsteins Erkenntnis zur neuen Ära der Astrophysik* ("Signals of Gravity: Gravitational Waves—Einstein's Discovery and the New Era of Astrophysics") (Stuttgart, Germany: Kosmos, 2012).

Vaas, Rüdiger, *Tunnel durch Raum und Zeit: Von Einstein zu Hawking: Schwarze Löcher, Zeitreisen und Überlichtgeschwindigkeit* ("A Tunnel Through Space and Time: From Einstein to Hawking—Black Holes, Time Travel and Superluminal Velocity") (Stuttgart, Germany: Kosmos, 2012).

Various Astronomical Topics

Ferreira, Pedro G., *The Perfect Theory: A Century of Geniuses and the Battle Over General Relativity* (New York: Houghton Mifflin Harcourt, 2014).

Frebel, Anna, *Auf der Suche nach den ältesten Sternen* ("The Search for the Oldest Stars") (Frankfurt am Main, Germany: S. Fischer Verlag, 2012).

Freistetter, Florian, *Asteroid Now: Warum die Zukunft der Menschheit in den Sternen liegt* ("Asteroid Now: Why the Future of Humanity Lies in the Stars") (Munich: Hanser Verlag, 2015).

Gates, Evalyn, *Einstein's Telescope: The Hunt for Dark Matter and Dark Energy in the Universe* (New York: W. W. Norton, 2009).

Jayawardhana, Ray, *The Neutrino Hunters: The Chase for the Ghost Particle and the Secrets of the Universe* (New York: Scientific American/Farrar, Straus & Giroux, 2013).

Panek, Richard, *The 4 Percent Universe: Dark Matter, Dark Energy, and the Race to Discover the Rest of Reality* (New York: Houghton Mifflin Harcourt, 2011).

Panek, Richard, *Seeing and Believing: How the Telescope Opened Our Eyes and Minds to the Heavens* (New York: Viking Penguin, 1998).

Perryman, Michael, *The Making of History's Greatest Star Map* (Berlin/Heidelberg: Springer, 2010).

Zimmerman, Robert, *The Universe in a Mirror: The Saga of the Hubble Space Telescope and the Visionaries Who Built It* (Princeton, NJ: Princeton University Press, 2008).

Image Credits

Index

Page numbers followed by *i* indicate insert photograph.

About the Author

FLORIAN FREISTETTER is an astronomer, author, columnist, blogger, and podcaster. Born in 1977, he studied astronomy at the University of Vienna. In 2008, he launched the astronomy blog *Astrodicticum Simplex,* one of the most widely read science blogs in German. His podcast, *Sternengeschichten* ("Star Stories"), is one of the most successful German-language science podcasts. His books include *Der Komet im Cocktailglass* ("The Comet in a Cocktail Glass," 2013), which was awarded the Austrian Science Book of the Year prize in 2014; *Isaac Newton: The Asshole Who Reinvented the Universe*; and most recently, *Stephen Hawking: His Science in a Nutshell,* among many others. In 2015, he became a permanent member of the Austrian "scientific cabaret," Science Busters. In 2013, the asteroid 2007 HT3 was officially designated 243073 Freistetter by the Minor Planet Center of the International Astronomical Union.